T0275641

CAMBRIDGE LIBRARY COLLECTION

Books of enduring scholarly value

Earth Sciences

In the nineteenth century, geology emerged as a distinct academic discipline. It pointed the way towards the theory of evolution, as scientists including Gideon Mantell, Adam Sedgwick, Charles Lyell and Roderick Murchison began to use the evidence of minerals, rock formations and fossils to demonstrate that the earth was older by millions of years than the conventional, Bible-based wisdom had supposed. They argued convincingly that the climate, flora and fauna of the distant past could be deduced from geological evidence. Volcanic activity, the formation of mountains, and the action of glaciers and rivers, tides and ocean currents also became better understood. This series includes landmark publications by pioneers of the modern earth sciences, who advanced the scientific understanding of our planet and the processes by which it is constantly re-shaped.

Observations in Meteorology

Although devoted to his parish, Leonard Jenyns (1800–93) combined his clerical duties with keen research into the natural world around him. His numerous publications include *A Manual of British Vertebrate Animals* (1835) and *Observations in Natural History* (1846), both of which are reissued in this series. This 1858 work is based on nineteen years of meticulous observation of Cambridgeshire weather, including trends in atmospheric pressure and precipitation. Jenyns' careful recording of his surroundings supplies the raw data for the text and many informative tables. The geological position of Swaffam Bulbeck, where most of the observations were made, is briefly discussed along with other factors bearing upon the climate of Cambridgeshire more generally. Throwing light on how meteorological observation was conducted and interpreted, the work reflects a growing interest in the topic in Victorian Britain.

Cambridge University Press has long been a pioneer in the reissuing of out-of-print titles from its own backlist, producing digital reprints of books that are still sought after by scholars and students but could not be reprinted economically using traditional technology. The Cambridge Library Collection extends this activity to a wider range of books which are still of importance to researchers and professionals, either for the source material they contain, or as landmarks in the history of their academic discipline.

Drawing from the world-renowned collections in the Cambridge University Library and other partner libraries, and guided by the advice of experts in each subject area, Cambridge University Press is using state-of-the-art scanning machines in its own Printing House to capture the content of each book selected for inclusion. The files are processed to give a consistently clear, crisp image, and the books finished to the high quality standard for which the Press is recognised around the world. The latest print-on-demand technology ensures that the books will remain available indefinitely, and that orders for single or multiple copies can quickly be supplied.

The Cambridge Library Collection brings back to life books of enduring scholarly value (including out-of-copyright works originally issued by other publishers) across a wide range of disciplines in the humanities and social sciences and in science and technology.

Observations in Meteorology

*Relating to Temperature,
the Winds, Atmospheric Pressure,
the Aqueous Phenomena of the Atmosphere,
Weather-Changes, Etc.*

LEONARD JENYNS

CAMBRIDGE
UNIVERSITY PRESS

CAMBRIDGE
UNIVERSITY PRESS

University Printing House, Cambridge, CB2 8BS, United Kingdom

Published in the United States of America by Cambridge University Press, New York

Cambridge University Press is part of the University of Cambridge.
It furthers the University's mission by disseminating knowledge in the pursuit of
education, learning and research at the highest international levels of excellence.

www.cambridge.org
Information on this title: www.cambridge.org/9781108069878

© in this compilation Cambridge University Press 2014

This edition first published 1858
This digitally printed version 2014

ISBN 978-1-108-06987-8 Paperback

OBSERVATIONS

IN

METEOROLOGY.

OBSERVATIONS

IN

METEOROLOGY:

RELATING TO

TEMPERATURE, THE WINDS, ATMOSPHERIC PRESSURE,
THE AQUEOUS PHENOMENA OF THE ATMOSPHERE,
WEATHER-CHANGES, ETC.,

BEING CHIEFLY

THE RESULTS OF A METEOROLOGICAL JOURNAL
KEPT FOR NINETEEN YEARS AT SWAFFHAM BULBECK,
IN CAMBRIDGESHIRE,

AND

SERVING AS A GUIDE TO THE CLIMATE OF
THAT PART OF ENGLAND.

BY

The Rev. LEONARD JENYNS, M.A., F.L.S., etc.

LATE VICAR OF SWAFFHAM BULBECK.

" When He uttereth his voice, there is a multitude of waters in the heavens, and
He causeth the vapours to ascend from the ends of the earth ; He maketh lightnings
with rain, and bringeth forth the wind out of his treasures."—*Jerem.* x. 13.

LONDON:

JOHN VAN VOORST, PATERNOSTER ROW.

MDCCCLVIII.

PRINTED BY TAYLOR AND FRANCIS,
RED LION COURT, FLEET STREET.

PREFACE.

THE Author of the present work has been an observer of the weather during a great part of his life. For a period of more than thirty years he kept a meteorological register, first at Bottisham Hall, and afterwards at Swaffham Bulbeck, the parish to which he was appointed in 1823. These two places are both in Cambridgeshire, and distant from each other scarcely half a mile. The observations he made were not at first conducted with sufficient regularity, or with sufficiently good instruments, to render the results of much value. But in the year 1831, desirous that his time and trouble should not be entirely thrown away, he commenced a more accurate system of observing, having previously obtained from Mr. Newman a barometer of the best construction that was to be had in those days, together with thermometers and a Daniell's hygrometer from the same maker. These instruments are further spoken of in the body of this work.

It was the Author's original intention to have

carried on this second set of observations for at least
twenty years from the date above mentioned, before
embodying the results, or attempting any generaliza-
tions respecting the weather. At the end of the nineteenth year, however, being called upon to quit that
neighbourhood, he was obliged from that time to dis-
continue the register; and in the hope of furnishing,
if possible, any contributions, however trifling, to the
science of Meteorology, which of late years has attracted so much more of the public attention than
formerly, as soon as circumstances permitted, he commenced putting together the results of his Journal as
they appear in the following work. These results
relate chiefly to temperature, the winds, the pressure
of the atmosphere, and the various phenomena of
clouds, dew, rain, snow, &c. It has been particularly
endeavoured to trace the connexion between different
states of weather, and the shiftings of the various
currents in the atmosphere, as indicated by the vane
and barometer jointly. He believes that so far as it is
possible to judge of the coming weather, and to determine to what degree the weather is to be trusted
at any particular time, it can only be done by close
attention to these points, and long familiarity with the
usual course of phenomena, attendant upon changes
from dry to wet, and *vice versâ*. At the same time
within what narrow limits any prognostications must
necessarily be confined, from our ignorance of the
weather prevailing at the same time in other places

than that in which the observer is situate, as well as
from various contingencies which no human foresight
can take knowledge of,—the Author has endeavoured
to point out in a chapter especially devoted to this
part of the subject.

Many of the results, more especially those relating
to the winds, and their ordinary course throughout the
year, have been compared with similar results given
by Howard in his ' Climate of London*,' which has
been often referred to as a standard in this respect ;
his work being perhaps, on the whole, the most complete we have on the climate of that part of England,
from which the climate of Cambridgeshire, excepting
so far as it is influenced by proximity to the fens,
does not materially differ.

There are certain parts of the subject of Meteorology, which, not having been particularly attended to
by the Author, are either entirely passed over, or but
slightly alluded to in this work. Such are the optical
and magnetic phenomena of the atmosphere, and
those connected with atmospherical electricity. Some
of these phenomena occur but seldom ; others are but
imperfectly understood : while from their relation to
higher departments of science than are studied by
most persons, they require less notice in a work intended to serve for general purposes.

It will be understood, therefore, that the present

* The Second Edition is the one referred to in this work, in
three vols. 8vo. Lond. 1833.

work is neither an elementary one, nor a complete treatise on any particular branches of Meteorology. It is simply an attempt to give a general character of the weather and seasons, as noticed in that part of England in which the observations have been made, and to point out the ordinary conditions under which weather-changes seem to take place. At the same time there are added a few observations made elsewhere, as well as much matter calculated to make the work equally useful to persons in other parts of the country, who, from their various out-of-door occupations, are naturally interested in the subject. In drawing up the whole, but especially the portions last alluded to, some assistance has been derived from different authorities, which, however, it is believed, in all cases, have been carefully mentioned.

Those who desire an introductory work, or a complete treatise on this department of science, may consult Thomson* or Kämtz†. Those who wish to have a practical acquaintance with the chief instruments used in making meteorological observations, are referred to a little work by Mr. Drew‡, in which they are fully described, with the addition of illustrative engravings, and various useful tables, &c.

It is accordingly hoped, that the work now offered

---

* *Introduction to Meteorology.* 8vo. Edinb. and Lond. 1849.

† *A complete Course of Meteorology.* (English translation by Walker.) 8vo. Lond. 1845.

‡ *Practical Meteorology.* 12mo. Lond. 1855.

to the public may be of more than local interest. It
is very probable, indeed, that some of the generaliza-
tions it contains, those especially respecting weather-
changes, and the circumstances attending them, may
prove erroneous, and based upon insufficient observation. The Author, while he trusts any errors of this
kind may be pardoned, begs to express his readiness
to be set right where wrong. It has simply been his
desire to make such deductions as the observations of
a moderately long series of years seemed to warrant,
and which must be always liable to correction from
the researches of others, particularly when these latter
have been carried on for still longer periods.

It may be stated, in conclusion, that Swaffham
Bulbeck, where the greater part of the observations
mentioned in this work were made, is situate between
seven and eight miles E.N.E. of Cambridge, and bordering upon the fenny parts of that county. Its estimated height above the level of the sea is rather more
than 20 feet. Its geological position is spoken of in
the last chapter, which contains a few general remarks
on the climate of the neighbourhood, though not intended to furnish all the elements for determining
the climate of Cambridgeshire at large.

Upper Swainswick, near Bath.
    Feb. 18th, 1858.

ERRATUM.

P. 24, 5 lines from the bottom, under the head of August, *instead of* 15°·3 *read* 16°·8.

TABLE OF CONTENTS.

OBSERVATIONS IN METEOROLOGY.

CHAPTER I.

THERMOMETER and TEMPERATURE.

(1.) The chief peculiarities of climate, as well as the different changes of weather, in a given locality, seem to depend primarily upon temperature. It is this, probably, which causes the different winds, and most of the phenomena of which Meteorology takes cognizance. Hence the Thermometer, the instrument employed in measuring the temperature, is in the hands of all persons who take the slightest interest in the subject. It is not my intention here to describe the different forms of this instrument in common use, or the principles of their construction, which may be found in various other works, but it may be well to say a few words respecting the method of fixing a thermometer. Great care is required in placing it, so as to ensure its giving a correct indication of the temperature of the air. For this purpose it is necessary, not merely to place it in the shade, facing the north, but to see that there be no wall, or other large object near, from which the sun's

B

rays can be reflected upon the instrument. It is further essential that the instrument be not itself fixed to a wall, the temperature of which it would take, and which is very slow in following the changes of the atmosphere. The best method perhaps is to fix it to a post or board, having a north wall a few feet behind, the instrument being about 5 feet from the ground: it must be exposed to a free current of air, but screened from rain and radiation upwards to the sky. Inattention to these points causes great differences in thermometers, though placed very near each other, and sometimes occasions much surprise with those who are not aware of the causes to which the differences may be traced. I have known two instruments, which stood alike when placed under exactly the same circumstances, show a difference of three or four degrees when the position of one was a little altered, so as to cause it to be influenced by radiation, from the effects of which the other was sheltered: yet these two thermometers were scarcely more than a few inches distant from each other.

(2.) This will serve to account for the extremely high and low temperatures which are sometimes recorded in the public prints as having occurred in particular localities, but which are to be received with caution, when no mention is made of the exact situation in which the instruments have been placed. When the weather is unusually hot or cold, many persons are induced by curiosity to notice the temperature, who seldom attend to it at other times; and not being aware of the great care required to ensure accuracy of observation, they often obtain results which are quite at variance with those of others, even in the same neighbourhood.

(3.) The instrument with which my observations were made at Swaffham Bulbeck, was one of Rutherford's self-registering thermometers, made by Newman. During the nineteen years over which the observations extended, neither the instrument nor the position of it was changed. This latter was much in accordance with what has been said above on this subject, excepting in elevation. The thermometer was fixed to a board, having a hood, projecting a short distance from the north side of a house, and about 10 feet from the ground. The sun never shone upon it, nor upon any near object from which a direct line could be drawn to the bulb of the instrument.

(4.) This thermometer was registered daily at 10 A.M. and 10 P.M., these hours being selected principally from motives of convenience. It is probable, however, that the mean of observations made at these hours is not very far removed from the mean temperature of the day*. The maximum and minimum temperatures during the twenty-four hours were also registered, and it is from the mean of these that the mean temperature

* According to Professor Dewey, the result obtained at the homonymous hours of 10 A.M. and 10 P.M. does not differ from the true mean diurnal temperature more than $\frac{5}{100}$ths of a degree. (See *Ed. Phil. Journ.* vol. vi. p. 352.) Dr. Brewster, in his memoir on the hourly observations made at Leith Fort (*Ed. Journ. of Sci.* vol. v.), considers the difference as amounting to ·122°. From these latter observations it appears that the exact periods of the day at which the mean temperature of the twenty-four hours occurs, are 9^h 13^m A.M., and 8^h 27^m P.M. But Dr. Brewster remarks that a single observation made every day at one of these two hours, more especially that at 9^h 13^m A.M., will give the mean temperature of the year with a great degree of accuracy.

for the whole period of the nineteen years has been calculated.   There have been many blank days, and occasionally whole weeks, during that period, when absence
from home, sickness and other causes, prevented any
observations being made.   It is not pretended, therefore, that the series is perfect, or that the deduced
results are more than an approximation to the truth.
It is only guaranteed that the observations, whenever
made, have been made with care, at the same times of
the day in all cases, and, as already stated, without any
alteration in the circumstances or position of the instrument; and it is only desired that they be received
for just so much as they are worth under the conditions.

(5.) The mean temperature of a place is one of the
first points to be ascertained in respect of its climate;
and however this may differ from that of other places,
even on the same parallel of latitude, it is remarkable
how little it varies at the same place, when the mean
temperatures of long terms of years are compared with
each other.   Howard, after having obtained the mean
temperature of certain stations in and near London
from the observations of many years, found that the
addition of fourteen years' further observation of the
temperature, at one of the stations, did not oblige him
to displace a single result*.   The mean temperature of
Greenwich for twenty years, commencing with 1810 and
ending with 1829, differs by less than half a degree
from the mean temperature of the twenty years next
succeeding†.   My own observations at Swaffham Bul-

---

* *Climate of London*, 2nd edit. vol. i. p. 1.

† This is deduced from Mr. Glaisher's Tables in the *Phil.
Trans.* for 1850, p. 589.

beck have not been sufficiently extended to admit of
any certain deductions on this subject; nevertheless I
find that, on dividing the period of time over which
they have been made into two equal portions, there is
very little more than half a degree difference between
the mean temperatures of the two series of years so
taken. There seems no good ground for supposing
that there has been any appreciable change in our
climate within the period of which we have historical
records. If such change has been observed in other
countries, or if, as some think*, it has taken place in
our own, it is not due, as Arago has clearly shown†, to
any cosmical causes, but to circumstances entirely local;
such as the clearing of woods, the draining of extensive
marshes, &c. At the same time there can be no doubt,
as Mr. Glaisher has observed‡, "that causes exist at
different times, which raise or depress the temperature,
and which continue through long periods." This, however, is quite distinct from any permanent alteration of
climate. The remark, too, appears intended to apply
exclusively to fluctuations of the mean temperature for
series of years, which is independent of fluctuations of
the particular seasons. The winters and summers might,
or might not, be colder and hotter than formerly, while
the mean temperature remained stationary, and *vice
versâ*. Whether the former *are* different or not, is an
inquiry worth making, but not one to be decided by
the loose recollections of persons advanced in life. It

* See a paper by Andrew Knight in *Hort. Trans.* vol. vii.
p. 563.

† "On the Thermometrical State of the Terrestrial Globe,"
Edinb. New Phil. Journ. vol. xvi. p. 205.

‡ *Phil. Trans.* 1850, p. 585.

is very common to hear reference made to what are
called old-fashioned winters and summers, whenever we
happen to have a season rather colder or hotter than
usual. But the individuals who speak in this way have
rarely made any accurate observations in times past to
allow of comparison with the time present. It is also
often owing to some association of ideas in their own
minds, that persons in general are led to think the
weather in certain years and seasons more extraordinary than it really is; while the weather at other
times, though in fact nearly similar, for want of this
association, fails to make any particular impression,
or to obtain lasting hold of the memory.

(6.) One circumstance which has very much led to
the opinions above stated respecting a supposed change
of climate in this country, is the fact of there being in
many parts of England, especially in the neighbourhood
of old towns, certain localities which bear the name of
vineyards. It has been thought that this indicates in
former times the cultivation of the grape for the making
of wine, for which our climate is now considered too
cold. But this argument is hardly sufficient to establish the point at issue. For though no doubt wine
has been formerly made in such places, causing them to
be called by the above name, it was long ago suggested
by Daines Barrington in the 'Archæologia*,' and the
suggestion was supported by many strong authorities,
that it might not have been wine from *grapes*, but from
currants, or some other fruit especially cultivated for the
purpose, which was chiefly made and drunk in those

* See his memoir on the growth of the vine in England, and
on "the Vineyard Controversy " (in reply to a previous memoir
by the Rev. Samuel Pegge) :—*Archæologia*, vol. iii. p. 67.

days. And just as we now make such wines, and still call them *wines*, though the produce of the *vine* is not concerned in the preparation of them, so formerly gardens in which the fruits were grown may have been called *vineyards*, from the purpose to which the produce was applied, though the true vine was not grown in them at all. But without having recourse to this explanation, I may state, in reference to a supposed alteration of our climate to affect the cultivation of the vine, that I have seen almost as fine grapes ripened in the open air in the south of England as any that I ever tasted grown in a hothouse. Such undoubtedly could not be obtained every season; but still, without knowing more exactly what the nature and quality of the wine really was which our ancestors were in the habit of making, and which may have been such as *we* should not call drinkable*, we can hardly speculate upon the conditions of the climate in those days in which such wine was made.

(7.) I will now give the results of my observations on the temperature of Swaffham Bulbeck in the following Table, which shows the mean temperature, as well as the extremes and range of temperature, in each year of the series, commencing with 1831 and ending with

* Humboldt observes that " the cultivation of the vine, to produce *drinkable* wine, requires not only a mean annual temperature of above $9\frac{1}{2}°$ Cent. (or 49°·2 Fahr.), but also a winter temperature of above 0°·5 C. (32°·8 F.), followed by a mean summer temperature of at least 18° C. (64°·4 F.)"—*Cosmos* (Sabine's Translation), vol. i. p. 322.

Mr. Glaisher sets the mean annual temperature of Greenwich, from the observations of seventy-nine years, at 48°·29 F.; that of winter at 37°·6, and that of summer at 60°.—*Phil. Trans.* 1850, p. 594.

8 OBSERVATIONS IN METEOROLOGY.

1849. The mean temperature in this Table is deduced from the extremes of each day.

TABLE I.—*Showing the mean, maximum, and minimum temperatures, and the range of temperature, during each of nineteen years, commencing with 1831 and ending with 1849.*

Year.	Thermometer.			
	Mean.	Maximum.	Minimum.	Range.
1831.	50·60	80	22	58
1832.	49·50	79·5	22	57·5
1833.	48·10	83	19	64
1834.	49·80	85	24·5	60·5
1835.	48·90	84·5	19	65·5
1836.	48·30	84	16·5	67·5
1837.	47·80	82	11	71
1838.	47·10	80·5	10	70·5
1839.	47·90	79	17	62
1840.	48·00	82	18	64
1841.	49·30	83	6	77
1842.	50·40	88	19·5	68·5
1843.	49·85	83	18	65
1844.	48·49	83	17·5	65·5
1845.	47·53	82	7	75
1846.	51·44	89	15	74
1847.	49·24	84	14	70
1848.	49·90	83	18	65
1849.	49·48	81·7	16	65·7
Means...	49·03	82·9	16·3	66·6

(8.) The year of highest mean temperature in the above series was 1846, when the mean rose to 51°·44. The lowest was 1838, when it fell to 47°·10. The difference between these two, equalling 4·34 degrees, shows the entire range of the mean during the whole period of observation.

(9.) The average mean is 49°·03, or, neglecting the second place of decimals, it may be set at 49°. This, as already stated, is deduced from the daily extremes. The same mean, calculated from the ten-o'clock morning and evening observations, is 48°·5, or half a degree less. The former, however, is probably the more correct of the two, from the circumstance of the morning and evening observations having been registered with less regularity than the daily extremes. Assuming 49° to be the true mean, it is singular that it should be exactly coincident with the mean temperature of Greenwich for the same series of years, as deduced from Mr. Glaisher's Tables in the Philosophical Transactions before referred to. If this last, however, be determined by a Table of temperatures at Greenwich, given by Mr. Belville in his 'Manual of the Thermometer*,' it is found to be 49°·27, or slightly higher than that of Swaffham Bulbeck.

(10.) I am not able, as I had wished, to compare the mean temperature of Swaffham Bulbeck with the mean temperature for the same period at the Cambridge Observatory, as, till of late years, meteorological observations have not been carried on there with much regularity. But taking the mean of certain particular years in the series, amounting to eight in number, for which I have obtained the mean temperatures, partly from the registers formerly published in the *Cambridge Chronicle*, and partly through the kindness of Professor Challis,

---

* P. 53. These temperatures appear to have been obtained from observations made elsewhere than at the Royal Observatory, as they do not accord with those published by Mr. Glaisher in the Philosophical Transactions for 1850. I refer to them exclusively in what follows, as the registering of the extreme daily temperatures at the Observatory did not commence till 1841.

10   OBSERVATIONS IN METEOROLOGY.

and comparing it with the mean of the *same years* at
Swaffham Bulbeck, there is very nearly a coincidence
here also, the mean temperature at the latter place being
one-tenth of a degree *lower* than at the Observatory.
(11.) The highest temperature registered at Swaffham
Bulbeck during the above nineteen years was 89°. This
occurred in 1846, the same year as that of the highest
mean. The lowest temperature was 6°, in 1841. Sub-
tracting the extreme minimum from the extreme maximum, the difference, amounting to 83°, shows the range
of the temperature during the whole period.
(12.) Neither of these extremes reaches those registered at Greenwich during the same period; the highest
maximum in Mr. Belville's Table being 90°·2, and the
lowest −4°. This last, therefore, presents the greatest
difference. But it is certain that local circumstances
will often occasion a much greater degree of cold at
one place than another, even when the two places are
considerably nearer together than those in question.
The year in which this low temperature occurred at
Greenwich was 1838, the minimum at Swaffham Bulbeck that same year not being below 10°; and that this
latter observation was probably correct, may be inferred
from the circumstance of the lowest temperature at
the Cambridge Observatory on the same occasion having
been only 10°·8, which presents but little difference.
It is observable, again, that in 1841, the year in which
the extreme minimum of 6° occurred at Swaffham
Bulbeck, the minimum at Greenwich was 5°·4, or very
little more than half a degree lower. Hence it must
have been quite accidental that, in the instance of the
year above mentioned, the difference between the minimum at Greenwich and that at Swaffham Bulbeck should

have amounted, as it did, to 14°. In no other year do I find a difference of more than 5°, and ordinarily it is much less.

(13.) But as a further proof of the remarkable difference in the degree of cold sometimes experienced in two places, even when contiguous, reference may be made to an instance mentioned by the Astronomer Royal, in a communication to Mr. Lowe of Nottingham, "On the Winter of 1844-45 at the Royal Observatory, Greenwich*." That winter was very severe in the months of December and February, the intermediate month of January being comparatively mild. The second frost began on the 25th of this last month, and continued with little interruption between seven and eight weeks. The coldest day in February was the 12th. On that day the temperature at the Greenwich Observatory was 7°·9, while at Lewisham, distant not quite a mile, it fell to $-1°·5$, or nearly $9\frac{1}{2}$ degrees lower than at the Observatory. At Swaffham Bulbeck, the temperature on that same day was not lower than 13°. This winter was remarkable for extending far into March, the first three weeks of which were as cold as February; and on the morning of the 14th of March, the thermometer at Swaffham Bulbeck actually fell to 7°, a greater depression of temperature than, perhaps, had been ever known before in this country, so near the time of the vernal equinox. And here, too, was another instance of much severer cold in one place than another distant only a few miles off. For at the Cambridge Observatory on that same day, the thermometer was not lower than 12° (according to observations published at the time by Mr. Berry, in the *Cambridge*

* Published by Mr. Lowe in his *Atmospheric Phenomena*, p. 322.

Chronicle), being 5° short of what it fell to at Swaffham Bulbeck.

(14.) These differences of temperature, on certain occasions, at different places not very far apart, are probably due to the air being, from accidental causes, relatively more free from mist and vapour, as well as more still in some spots than others, thereby favouring a greater intensity of terrestrial radiation; something is due also to difference of elevation. It is likewise necessary to know, in all cases of very low temperatures, how thermometers are placed, and how high above the ground; for very slight variations in the mode of fixing the instruments may materially affect the results. Thus, in the instance of the 20th of January, 1838, when my thermometer, wholly sheltered from radiation, fell to 10°, I found that another instrument *close by*, but not quite so much screened, was as low as 8°, a difference of 2 degrees. This shows the precautions necessary to ensure accuracy, as noticed at the beginning of this chapter.

(15.) Though the two extremes of temperature above given are the highest and lowest that occurred at Swaffham Bulbeck during the nineteen years before stated, they are not the widest extremes I ever registered in that neighbourhood. In July 1825, the thermometer three times rose above 90°, standing on the 19th of that month as high as 93°*; while on the morning of the 15th of January 1820, it descended to 5°. These extremes give a range of temperature during twenty-nine years of 88°, and the mean of the two being 49°,

* The day previous, the 18th, the thermometer stood at Datchet, near Windsor, as high as 96°. See *Phil. Trans.* 1826, p. 70.

it is curious to observe how nearly this approaches the mean temperature, as calculated from the daily extremes already spoken of. The mean of the *yearly* extremes of the series, commencing with 1831, and ending with 1849, also approaches that of the daily extremes for the same period within very little more than half a degree, and no doubt is more to be depended on in the long run. It is probable, indeed, that the mean obtained in this way, if deduced from the observations of a very long term of years, would be found almost exactly coincident with the true mean. And this gives a peculiar value to the self-registering thermometer, which might be availed of to get the mean temperature of places in which observations could not be made with any frequency or regularity. Howard has a remark on this subject: he observes, "it is possible, that in the thermometer of Six we possess an instrument, which being merely fixed to a post, and properly defended from the sun's rays and from accidents, in an uninhabited country, where it could be visited and adjusted by navigators once in every year, would give in a moderate run of years, with considerable accuracy, the mean temperature of the latitude and elevation where it stood *."

(16.) The temperature of 96°, registered at Datchet on the 18th of July, 1825, as stated in the last note, is, so far as I am aware, the highest on record in this country. Howard, however, had once before observed the same degree of heat. This last instance was on the 13th of July, 1808, on which day, according to Mr. Belville, the *mean* temperature of the twenty-four hours at Greenwich was nearly 80°, " being equal," he observes,

* *Climate of London*, vol. i. p. 19.

"to the heat of Egypt or Upper Ethiopia*." With regard to the greatest degree of cold, Mr. Glaisher states, that, on the 24th of December, 1796, in the environs of London the thermometer fell to $-6°$†. A lower temperature than this has probably not occurred since in the southern half of the kingdom ‡; though on the 9th of February, 1816, according to Howard §, the temperature fell to $-5°$. Assuming $96°$ and $-6°$ as the extremes of the climate, we have $102°$ as the whole range of temperature to which this part of England is subject.

(17.) Howard has observed, that "of the extremes of *cold*, the far greater number occur in January, only two" (in his series of ten years, commencing with 1807, and ending with 1816) "being in December and one in February. The extremes of *heat* are more diffused : only five of them fall in July, and the remainder in diminishing proportion earlier and later in the summer. Thus of the whole twelve, there are only two months in spring, and two in autumn which are not occasionally subject to one or the other annual extreme of temperature∥." In my series of nineteen years, the *lowest* temperature occurred eight times in January, five times in December, and three times in each of the months of February and March : the *highest* temperature occurred eight times in July, five times in August, three times in June, twice in May, and once in September. Here, therefore, the only months exempted

---

* *Man. of Therm.* p. 44.   † *Phil. Trans.* 1850, p. 603.

‡ At Alford, in Aberdeenshire, the thermometer is said to have fallen to $-12°$ on the 15th of Feb. 1838. See *Edinb. New Phil. Journ.*, vol. xxviii. p. 322.

§ *Climate*, &c., vol. i. p. 21.   ∥ Ibid. p. 18.

from either of the two annual extremes were those of April, October, and November.

(18.) The *mean* yearly maximum in my series I find to be $82°\cdot9$: that at Greenwich for the same series of years is $83°\cdot8$. The lowest maximum that occurred in my series was $79°$, in 1839; this year, and 1832, in which the maximum was half a degree higher, are the only years in which it did not rise to $80°$. There are but two years in the series, 1842 and 1846, in which it rose above $85°$. The average number of days in the year in which the temperature may rise to or above $80°$, is set by Belville at $6\cdot5$: I find it about 7.

(19.) The *mean* yearly minimum in the series is $16°\cdot3$, that at Greenwich is $15°\cdot5$. The highest is $24°\cdot5$. There are but three years in which the minimum did not descend below $20°$: these years were 1831, 1832, and 1834. There are likewise but two years, 1841 and 1845, in which it fell below $10°$.

(20.) The *mean* yearly range is $66°\cdot5$: the greatest in any year is $77°$, and the least $57°\cdot5$. The mean yearly range for the same series of years at Greenwich is $68°\cdot3$.

(21.) Thus, on the whole, it appears that (as determined by Mr. Belville's Tables) the mean temperature at Greenwich for the series of years in question was $0°\cdot24$ higher than at Swaffham Bulbeck. The average yearly maximum was likewise $0°\cdot9$ higher : the average yearly minimum $0°\cdot8$ lower : the average yearly range $1°\cdot7$ higher.

(22.) I have compared the above results with those obtained at Greenwich, the latter being more to be depended upon as correct than any others to which I could refer for the same series of years. The differences between them are not very considerable; and probably

if my own observations had been more complete, there
would be a nearer approximation.

(23.) From the consideration of the mean yearly
temperature, and yearly extremes, at Swaffham Bulbeck, I pass on to the temperature of each month in
the year taken separately, deduced from the same series
of nineteen years. I first give the average, highest and
lowest mean, and the range of the mean.

TABLE II.—*Showing the mean temperature of each
month in the year, deduced from nineteen years' observations, together with the highest and lowest mean,
and the range of the mean, during the same series of
years.*

Month.	Thermometer.			
	Mean.	Highest mean.	Lowest mean.	Range of mean.
January	36°·88	44°·70	29°·30	15°·40
February	38·25	43·98	31·30	12·68
March	41·70	46·80	34·83	11·97
April........	45·88	50·20	40·80	9·40
May	53·14	60·20	48·00	12·20
June	59·25	65·18	54·50	10·68
July	62·20	65·65	58·60	7·05
August......	61·74	67·20	58·29	8·91
September ...	56·94	60·50	52·06	8·44
October	50·17	56·20	46·00	10·20
November ...	42·67	45·57	38·60	6·97
December ...	39·19	44·60	31·21	13·39

(24.) By the above Table, it appears that, on an
average, January is the coldest, and July the hottest
month in the year; but taking the nineteen years separately, I find actually that the coldest month has occurred in January eight times, in December and February each five times, and in March once. The hottest

month has occurred in July ten times, in August seven times, in June once, and in May once.

(25.) Some curious and interesting results offer themselves, with respect to the possible relative temperatures of the different months, if we compare the extreme temperatures of each month with the extremes of all the rest. Thus it appears that—

January, though usually the coldest month, may be milder than either February, March or April. It is also remarkable that a higher mean has been attained in January than in February.

February may be milder than either March or April.

March may be warmer than April, but never attains the temperature of May.

April may be warmer than May, but never attains the temperature of June. On the other hand, it is observable that the lowest mean temperature of April is exceeded by the highest mean of all the other months of the year; so we might suppose it possible that in certain years it might be the coldest month of any, though in fact this probably never occurs.

In like manner the highest mean of May exceeds the lowest mean of June, July and August, though a year seldom occurs in which it is really the hottest month. It is a singular circumstance, and a striking instance of the strange irregularity of an English climate, *that two months should be thus next each other, one of which sometimes has the same temperature as that which is ordinarily the coldest month in the year, and the other sometimes the same temperature as that which is ordinarily the hottest.*

June, July and August, generally the three hottest months, may, any one of them, be colder than September.

September may be colder than October, but October is never colder than November.

November may be 6° colder than December, and December to a still greater extent (nearly 13½ degrees) colder than January.

(26.) If we look to the *average* mean of each month, we find the temperature gradually increasing from February to July, but not making equal advances from month to month. The *increment* becomes greater with each month till May, when it attains its maximum; that month being 7° warmer than April: after May it declines. In like manner, from August onwards to November, there is an increasing *decrement* of temperature, the decrement attaining its maximum in the latter of those two months, which, as it were, answers to May in this respect. In December this decrement declines, and it becomes still less in January, when it falls to a minimum.

(27.) The range of the mean is also very different in different months. On the whole, it is found to be considerably greater in winter than in summer. The month in which it attains its maximum is January. Its minimum would seem to be in November, but it is scarcely higher in July.

(28.) It has often been a question with meteorologists which month in the year, in respect of its mean temperature, most nearly represents that of the whole year. By Kirwan it was thought to be April; by Humboldt October. Some have stated that in *excessive* climates the mean temperature of April is generally as high as that of the year, while that of October is considerably higher; and in *moderate* climes that the former is generally as much lower as the latter is higher.

(29.) My own observations, as seen in the above Table, would go to show that neither month exactly represents the mean temperature of the year, that of April being too low by about 3°, and that of October too high by rather more than 1°; but that October approaches it most nearly. Taking the years of the series separately, March was found to be the month in one instance, and May in two instances. In all the other years it was either April or October.

(30.) From the subject of the mean temperature of the several months I proceed to that of the extremes and range of temperature of each month respectively, as exhibited in the following Table.

TABLE III.—*Showing the maximum and minimum temperatures for every month in the year; including the highest and lowest, as well as the mean and range, of each extreme; also the mean range of the temperature in each month, or the difference between the mean maximum and the mean minimum.*

Month.	Maxima.			Minima.			Range.		
	Mean max.	Highest max.	Lowest max.	Mean min.	Highest min.	Lowest min.	Range of the max.	Range of the min.	Mean range of each month.
January ...	51·04	57·5	43	19·58	28	6	14·5	22	31·46
February .	53·58	61·5	47·5	22·81	31	13	14	18	30·77
March......	58·71	66	49	23·66	31	7	17	24	35·05
April	67·82	76·5	59	27·86	31·5	23	17·5	8·5	39·96
May	74·00	83·5	65·5	33·75	38	26	18	12	40·25
June	78·23	86	70	41·66	46	35	16	11	36·57
July	80·61	89	71	42·79	48	36	18	12	37·82
August ...	79·47	88	71	43·13	50	38	17	12	36·34
September.	74·53	80·5	67·5	36·35	43	28	13	15	38·18
October ...	66·08	74	61	31·42	37·5	26	13	11·5	34·66
November .	57·92	62	54	26·79	30·5	24	8	6·5	31·13
December .	54·40	58	44·5	24·23	31·5	15	13·5	16·5	30·17

(31.) It appears from the above Table, in reference to the highest temperature attained in each month during the series, that there are five months, May, June, July, August and September, in which it occasionally rises to 80° or upwards, and only two, January and December, in which it never attains to 60°.

(32.) In reference to the lowest temperature, there are but three months, June, July and August, in which it does not occasionally descend to freezing-point, causing at times even sharp frosts, without taking in the effects of terrestrial radiation, to be hereafter spoken of. If we look, however, to the *mean* minimum, we may add May and September to the number of months ordinarily exempted from frost. Looking to the *highest* minimum attained in each month, October is the only month besides which has been ever known to be without frost, while in no instance this has been the case with April, the corresponding month in spring.

(33.) The mean range of temperature in each month appears to attain a maximum in April and May. It is less by several degrees in June, July, August and September, and remains nearly the same during the whole of that period. In October it has a further fall. During the next four months it descends to a minimum, undergoing, as before, little variation. It then rises in March to about what it was in October, before getting to the maximum again.

(34.) I proceed to consider the temperature of each season, including under the name of Spring, the months of March, April and May; under that of Summer, June, July and August; under that of Autumn, September, October and November; under that of Winter, December, January and February.

Table IV.—*Showing the mean temperature of each season, deduced from the observations of nineteen years,
with the highest and lowest mean, and the range of the
mean during that period.*

Season.	Mean temp.	Highest mean.	Lowest mean.	Range of the mean.
Spring	47·18	50·05	41·63	8·42
Summer.....	60·87	65·88	59·60	6·28
Autumn	49·86	52·07	48·67	3·40
Winter	38·09	42·93	32·73	10·20

(35.) It will be seen by the above Table, that—

The difference between the mean temperature of Spring
and that of Summer .... $= 13·69$
Ditto between the mean temperature of Summer and
that of Autumn .... $= 11·01$
Ditto between the mean temperature of Autumn and
that of Winter .... $= 11·77$
Ditto between the mean temperature of Winter and that
of Spring.................................. $= 9·09$

Thus the temperature, after the winter is over, advances more quickly as the year advances; that is to
say, summer follows more quickly upon spring, than
spring does upon winter. In fact the difference in
temperature between the last two seasons is less than
that between any two others, while that between spring
and summer is the greatest. The rate of decline of the
temperature in autumn is nearly a mean between the
two.

(36.) The range of the mean temperature of the respective seasons is greatest in winter; it becomes gradually less in spring and summer, and is least in
autumn, when it is only óne-third of what it is in

winter. From this it would seem that winter is a more variable season in this climate than any other, and that autumn is the one which preserves the most uniform character.

(37.) *Mean Daily Temperature.*—If there were no disturbing influences of any kind, the mean diurnal temperature would advance by regular steps each day from the coldest to the warmest period of the year, and afterwards decline in like manner. From various causes, however, some local, others merely temporary, this regularity is continually interrupted, so that the advance is seldom uniform for a week together. It is only after taking the mean of observations extending over a considerable period of years, that the chief variations arising from such causes disappear. Such has been done in a Table published by Mr. Belville*, in which is given the mean temperature for every day in the year, as observed at Greenwich, on an average of thirty-five years; and it is shown to advance, for the most part very evenly, from $34°\cdot22$, the lowest that occurs, falling on the 8th of January, to $63°\cdot65$, the highest, occurring on the 17th of July†. The day of the year, the mean temperature

---

* *Man. of Therm.* p. 50.

† A remarkable exception to the regularity of this advance has been observed on three days in May, viz. the 11th, 12th and 13th, on which there has been found (by eighty-six years of observations at Berlin) a *retrogression* of the mean daily temperature, equalling $2°\cdot2$ Fahr. It has been thought by Arago and others to be due to the intervention of asteroids between the earth and the sun at this time. See Humboldt's *Cosmos* (Sabine's translation), vol. i. p. 123, and note (86), p. 389. Arago states the break in the progressive increase of temperature as occupying the period included between the 5th and 13th of that month. The subject, however, of the " Determination of

of which, according to this table, approaches most nearly to the mean yearly temperature, is the 28th of April, the difference being merely one-tenth of a degree; so that we might infer, if observations with the thermometer were made on this day only, and carried over a very long period of years, they would be sufficient in the end to establish the mean temperature of the place in which they were made, with tolerable accuracy. Strictly speaking, however, this would apply only to Greenwich, where the above Table was constructed. A period of 110 days intervenes between the coldest day and the day of the mean, while there are only eighty days between the latter and the hottest. This accords with what was before observed respecting the more gradual advance of the temperature from winter to spring than from spring to summer (35). It is also noticeable in Mr. Belville's Table that there is not a single day in the year the mean temperature of which, taken on an average, falls below the freezing-point, though in certain seasons it may fall several degrees below it for weeks together.

(38.) *Diurnal Range.*—The diurnal range of the thermometer, or the difference between the maximum and minimum temperatures, is as variable as the daily mean; being dependent upon season, the direction of the wind, and the general state of the weather. Sometimes, when the wind remains steady in one quarter, and the air is still and the sky clouded, the tempe-

the Mean Temperature of every day " has more recently been investigated by Mr. Glaisher, who has noticed two or three other periods of the year in which there are interruptions of the regularity of the advance and decline of temperature, " very remarkable," he observes, " on account of the difficulty of assigning a physical cause." See *Athenæum*, 1857, p. 152.

rature scarce varies throughout the twenty-four hours, or it may even remain uniform for several days together. At other times, in bright weather, it will rise as much above the mean during the day as it falls below it at night; and the range may extend to 20, 30, or even near 40 degrees. The average range, in fine settled weather, taking one season with another, seems to be about 20 degrees. Occasionally, during winter, the maximum and minimum may be reversed in respect of the time when they occur; as on the sudden breaking up of a frost, when the temperature sometimes rises, in the course of two or three hours, 20 or 30 degrees : if this happen at night, the maximum may occur then; and if, as sometimes is the case, the frost return the next day, from the shifting of the wind, the thermometer may fall again, and lower than in the previous night; or there may be a double maximum and minimum from the temperature rising and falling twice during the twenty-four hours. These double maxima and minima should be taken into account, when the mean temperature is deduced from the daily extremes.

(39.) The following Table shows the mean diurnal range of the thermometer for each month in the year at Swaffham Bulbeck, on an average of nineteen years :—

Month.	Jan.	Feb.	Mar.	April.	May.	June.	July.	Aug.	Sept.	Oct.	Nov.	Dec.
Mean diurnal range .	8·1°	10·3°	13·2°	15·8°	19·6°	17·1°	17·6°	15·3°	15·7°	12·3°	10·6°	8·4°

(40.) The mean diurnal range for the season of spring is $15°\cdot7$; for that of summer $17°\cdot2$; for autumn $12°\cdot4$; for winter $9°$; being nearly double in summer what it is in this last season.

(41.) The mean diurnal range throughout the year

is 13·6.   Howard gives "the mean variation of temperature, from the heat of the day to the cold of the
night" (or, in fact, the mean diurnal range), in the
country about London, on an average of ten years, as
15°·4.   And he observes, what is interesting, that, in
London itself, upon the same average, it is only 11·37;
showing a difference of 4°·04.   The greater part of this
difference he states to be on the side of the *lower*
extreme, thus affording a proof how much warmer
London is than the country *at night* at all seasons of
the year*.

(42.) According to Howard, "the difference between
the mean temperatures of day and night coincides to a
fraction of a degree with the difference between those
of summer and winter†."

(43.) *Conditions of Extremes of Temperature.*—The
highest and lowest temperatures ever experienced in
this country are of rare occurrence : perhaps not twice,
or above twice, in a century does the thermometer rise
to 96°, or fall to −6°, the probable extremes of our
climate, as before mentioned.   It is interesting therefore to investigate the conditions under which such
unusual temperatures occur.   Howard has stated these
conditions, which, as they are quite confirmed by all I
have observed myself, I will repeat here.   He remarks,
that "to produce the highest possible temperature in
our climate, *or the extreme of heat*, there appears to
be required,—

"First, a *clear atmosphere* at the time; that the sun's
rays may have the freest possible access to the earth's
surface.

"Secondly, a *dry and warm state of the soil*, to some

* *Climate of London*, vol. i. pp. 28, 29.   † *Id.* p. 53.

considerable depth; that the earth may reverberate freely, without throwing up such a quantity of vapour as by its speedy condensation, in the higher and colder regions of the atmosphere, might produce cloudiness and annul the first condition.

" Thirdly, these two causes must concur at a season when the sun is not far from its greatest elevation: otherwise the heat will be in excess, only relatively to the time of year at which it occurs.

" Fourthly, to carry the heat to the very highest point, we must receive, at this crisis, by means of steady southerly breezes, the air of the southern parts of Europe; while these in their turn are supplied from Africa and the south of Asia.

(44.) " To produce the lowest possible temperature in our climate, or the *extreme of cold*, there is required,—

" First, as in the former case, a clear and dry atmosphere at the time, that the heat may freely escape by radiation.

" Secondly, a *cold state of the soil* (the usual result of previous cloudy, wet and frosty weather), and this to some considerable depth; that the sun's rays may not be assisted by any warmth from beneath in raising the temperature by day.

" Thirdly, the concurrence of these two causes with a sufficiently low state of the sun, and consequent length of night; otherwise the cold, although severe for the season, will not be such as to be remarkable in comparing together the results of a series of years.

" Fourthly, the winds must come to us from the northward; when, if they blow with sufficient steadiness, we may receive them at length from Siberia*."

* *Howard, Climate, &c.* i. p. 24.

(45.) These conditions have prevailed more or less on all the occasions of remarkable heat or cold I have myself noticed, though I never knew any extremes equal to those recorded by Howard. The hot days that occurred in July 1825, before alluded to, were attendant upon a very dry season, and accompanied by a wind varying from S. to S.E., though from the state of the barometer, which was above 30 inches, it is probable there was a higher current from the N. or N.E. So likewise in 1846, there had been a great drought of nearly five weeks in May and June, preceding the high temperature which occurred on the 5th of July: the wind on that day worked from W. to S.E. through the S., and then back again, the hottest period being coincident with its passage through the S. the second time. Though the thermometer at Swaffham Bulbeck on this occasion was not higher than 89°, it was stated in the *Cambridge Chronicle* to have been nearly 93° at the Cambridge Observatory.

(46.) In every instance in which the thermometer has been above 85°, during the whole series of nineteen years, commencing with 1831 and ending with 1849, the wind has been either S.E. or S.S.E.

(47.) Very low temperatures are generally accompanied by E. and N.E. winds, but occasionally N.W. This last was the prevailing wind on the 8th of January, 1841, which deserves mention as the very coldest day I ever remember, not merely in respect of the minimum, but also of the mean temperature of the twenty-four hours. There had been much frost and snow previously, especially during the latter half of December. On the morning of the day in question, at 7 o'clock, the thermometer was 7°, with thick fog and copious rime on the trees. By noon it had risen to 18°, the fog at that time beginning to dissipate, and the sun partially seen, with the air full of floating *spiculæ*: at 2 P.M. it had fallen again to 12°, the sky having been quite clear since noon, though the fog was now beginning to reform; it continued nearly stationary for a few hours, then fell gradually lower, till at 10 P.M., under a sky again clear and a bright moon, it fell to 6°*. The mean temperature of the twenty-four hours, included between midnight on the 7th of January and midnight on the 8th, deduced from eight observations, was only 11°, being more than 23° below the average mean tempe-- rature of that particular day in the year, as given in Mr. Belville's Table before alluded to (37).

(48.) A rather remarkable exception to Howard's third condition requisite for producing a very low tem- perature, occurred on the 14th of March, 1845, when, as before alluded to (13), the thermometer at Swaffham Bulbeck fell to 7°. Perhaps an equal degree of cold had never been registered before, so long after the period of the winter solstice. The weather, however, during the previous part of that month, as also during the whole of February, had been extremely severe. The wind had been generally E. or N.E. from the be- ginning of March; on the 11th and 12th it had shifted more to the N. and N.W., attended by a fall of snow which covered the ground to a considerable depth; on the 13th it was again easterly, and the sky quite clear of clouds, the temperature this day only $18\frac{1}{2}°$ in the morning, and never rising higher than $23\frac{1}{2}°$; on the 14th, at the period of the greatest cold, the wind was N.E., followed by another bright cloudless day,

* At Bottisham Hall, distant about half a mile, it fell to 5°.

but the thermometer, though only 7° a little before sunrise, got up to 33° during the day, and never afterwards descended lower than 19°, though the frost continued severe for a week afterwards. A phenomenon of this kind is one of those rare exceptions to a general rule, which hardly vitiates the rule itself, when taken in its ordinary operations.

(49.) *Coldest night before a thaw.*—I have often noticed that the coldest night during a frost of any continuance, is the night immediately preceding a thaw. This, though not always, is at least very frequently the case*. The circumstance seems capable of explanation in the following way. The coldest winds in winter are from the N.E., and these are the winds which most generally prevail in severe weather. The breaking up of a frost is attended by a change of wind to S.W., the passage being usually through S.E. and S. It is, however, in the higher regions of the atmosphere that this change commences; and it is due to the *first setting in* of a warmer current above from the S. or S.W.,—which, by its higher temperature, dissolves any cloudiness that may exist in the sky at that elevation, and so favours terrestrial radiation below, where the wind still remains unchanged,—that the cold is for a time increased. This effect, however, does not last long. As a larger body of warmer air flows in, cirrus and cumulus clouds soon make their appearance; the increase of temperature is gradually communicated to the lower strata of the atmosphere, these now following the course of the

* Belville has likewise mentioned instances, during three severe winters, in which " the *extreme* of cold was immediately succeeded by a change of wind to the S. and S.W., and that a thaw ensued."—*Man. of Therm.* p. 38.

upper, and thence to the ground, when the thaw commences *.

(50.) The following extracts from my Journal are a few selected instances of what has been just stated.

(Ex. 1.) Dec. 1846.—A very cold month, with little interruption of frost, at times severe : northerly winds prevailing.

Dec. 16. Min. temp. (of the night preceding) 20°. Wind N.

17.   „   „   26°.   „   N

18.   „   „   15°.   „   N.E.

but passing to S.W. (through S.E. and S.) in P.M. and followed by three days of very mild weather previous to a return of frost.

(Ex. 2.) March, 1847.—First half of the month cold and wintry; frost on and off.

March 10.—Min. temp. 26°. Wind N.E., passing to due E. in P.M., with a thick snow-storm at the time of passage : very sharp frost at night, with clear sky and rising barometer.

March 11.—Min. temp. 14°. Wind due E. at sunrise, but passing to S.W. in P.M. (through S.E. and S.), and followed by milder weather.

(Ex. 3.) Jan. 1848.—A seasonable winter month, frost setting in on the 6th, and continuing with only slight interruption occasionally till the 29th.

Jan. 25.—Min. temp. 30°. Wind N.E. Sky much clouded.

26.   „   27°·5.   „   E.   Less clouded.

27.   „   22°·5.   „   S.E. Fine and clear.

28.   „   18°.   „   S.E., passing to S. and

S.W. in P.M., thaw following.

(51.) We thus see how temperature is liable to be affected by other circumstances than the direction of the wind. Extreme cold is always the result of a clear

---

* The above explanation had in part suggested itself to me, before meeting with some remarks by Kämtz on temperature, in connexion with the state of the sky, and direction of the wind ; for which, see his *Meteorology* (by Walker), p. 166.

sky, without which north-easterly winds, in winter the coldest, have a higher temperature than naturally belongs to them. In the last of the above cases, the temperature on the 26th of January, with an easterly wind, and a sky partially clouded, was lower than on the 25th, when the wind was N.E., and the sky more thickly overcast : on the 27th, with the wind S.E., and a clear sky, it was lower than on the 26th ; while on the 28th, when the wind was still S.E., but about to change to S.W., it was lower than any other night during the frost.

(52.) *Extremes of Temperature in connexion with the Moon's changes.*—In the Report of the British Association for 1850*, Mr. R. Edmonds of Penzance has brought together some remarkable maxima of temperature, during a period of twelve years, commencing with 1839 and ending with 1850, all which occurred nearer to the moon's first quarter than to any other. This led me to inquire how far there might be any coincidence of the same kind in the series of years over which my own observations extended. It is hardly necessary to give the results in detail, as they appear to me of uncertain value. But so far as any conclusion can be drawn, it is favourable to the views of Mr. Edmonds. Thus, taking the yearly maxima for seventeen years, I find the numbers as follows in reference to the periods of the moon's four quarters :—

Number of maxima nearer to the first quarter ... 7.

  Ditto ,, ,, ,, second ditto ... 4.

  Ditto ,, ,, ,, third ditto ... 4.

  Ditto ,, ,, ,, fourth ditto ... 2.

This would seem to show a decided excess of yearly

* *Communications to the Sections*, p. 32.

maxima occurring nearer to the moon's first quarter
than to any other. If we take the number occurring
between the new and the full moon, as opposed to the
number *between the full and the new,* we find them in
the ratio of 10 to 7. But it is questionable whether
this may not be accidental, and whether in other instances the numbers might not be more nearly equal.

(53.) Not confining our attention to the yearly
maxima, but taking into account all the maxima above
80°, the result is somewhat similar; though not so
striking in respect of the number occurring nearer to
the first quarter, as of the number occurring between
the new and the full. It seems, on the whole, just to
warrant the conclusion, though it must be received
with caution, until further verified, that high temperatures occur more frequently between the new and the
full moon than between the full and the new.

(54.) With respect to the yearly and other *minima,*
I do not see that there is any connexion between the
times at which these occur and the moon's changes.
The number is nearly the same for all the quarters.

(55.) *Periods of greatest heat and cold.*—It has been
before stated that the days of highest and lowest *mean*
temperature in the year at Greenwich, in Mr. Belville's
Table, are the 17th of July and the 8th of January
respectively (37). These, however, are not necessarily coincident with the days of the yearly *maximum* and *minimum,* which in one sense may be called, and often are
called in common parlance, the hottest and coldest days
in the year. In most years, however, the former would
not be separated from the latter by any wide interval
of time. But it requires observations extended over
a very long series of years to fix these days with any

precision. They would also be different in different stations, like the days of extreme mean temperature, in consequence of local influences, connected with soil, configuration of the ground, and such like. For these reasons observers are not agreed in respect of the exact days. " According to Kämtz, the yearly maximum occurs on the 26th of July, and the minimum on the 14th of January. M. Crahay, from observations at Maestricht from 1818–33, gives the maximum upon the 19th of July, and the minimum on the 22nd of January. Bouvard found the maximum at Paris to be upon the 15th of July, and the minimum on the 14th of January. Arago, from very extensive data, obtained at the Paris Observatory between 1665–1823, found the day of greatest heat to oscillate irregularly in the months of July and August, while that of most intense cold usually fell about the middle of January *."

(56.) With respect to the periods of the *daily* maximum and minimum, or the hours in each day at which occur the greatest degrees of heat and cold, there is also some uncertainty, even at the same place, dependent upon the season and the character of the weather. Under a bright sky, and all other conditions the same, the period of the maximum seems to vary from half-past one in the afternoon, the hour in the middle of winter, to half-past two or three o'clock in summer. The minimum is generally considered as occurring about half-an-hour before sunrise. But this is not always the case, at least in winter. During very severe frost especially, I have watched the thermometer, and sometimes found it continuing to fall till sunrise itself, or even a few minutes after, without the slightest indication of

* *Thomson's Introduction to Meteorology*, p. 41.

c 5

any change in the state of the sky, or the direction of the wind, or the stillness of the atmosphere,—all circumstances, in fact, having remained precisely the same as they had been throughout the night.   Quetelet, referring to Brussels, considers the coldest time in the day as occurring about 6 A.M. in winter, and a quarter past three in summer.

(57.) *Haze before heat.*—A sudden increase of temperature, more especially on the first setting in of hot weather, is almost always preceded by a peculiar haze or mist, which first appears in the horizon, in the direction of the wind, attended by a slight breeze, and thence gradually diffuses itself through the air. Distant objects, which only a few minutes before were clear and well defined, assume an indistinctness of outline giving the appearance of the air being full of dust.   Sometimes this haze increases to such a degree as quite to render the sky overcast, and an ordinary observer might suppose that it was going to rain; but the great height of the barometer, which occasionally is even rising at the time, as well as all the other circumstances of the weather, indicate that there is no cause for any apprehensions of this nature. It is this state of the sky which often leads to the remark that it will either rain or *turn to heat.* After the current of heated air, which gives rise to the haze, has well mixed itself with the atmosphere of the place in which the observer is situated, the haze goes off, and the sky becomes more or less clear again, excepting that *cumulus* clouds generally form within a short time.

(58.) The following extract from my Journal will give an instance of this phenomenon, with the conditions of weather on the three previous days :—

1845. June 9.—Very fine and settled, with high barom.

(30·370) and westerly wind, after several days of strong S.W. wind, with cool clouded sky and occasional rain. Maxim. temp. 68°·5.

June 10.—As yesterday, but wind has worked to N.W.: barom. about the same: maxim. temp. 74°·5. *Cumulus* clouds yesterday; today only *cirrus*.

June 11.—Bright and cloudless. Wind N.E., having apparently worked on in the night from N.W. through the N. Barom. a trifle lower than yesterday: max. temp. 74°. Evening cool; sky still clear, and *very deep red sunset*.

June 12.—Brilliant summer's day, and very hot. Wind E.S.E. Barom. slightly fallen, but still standing at 30·195. Max. temp. 80°.—*Much haze all* A.M., *so that objects even at short distances appeared somewhat indistinct*: in P.M. cumuli began to form, the haze gradually disappearing. Evening quite cloudless, and very settled, with a temperature, at 10 o'clock, 7° higher than at the same hour yesterday evening.

On the two following days the temperature rose to 82° and 82°·5 respectively, with easterly wind, and the barom. nearly steady, not falling below 30·101.

(59.) In the above instance, it would seem to have been the current next the earth only that revolved from W. to N.E., the air above all the while, as well as previously, having flowed from this last quarter, or some point near it, as indicated by the high state of the barometer. As this lower current advances, the air becomes drier, the cumuli first giving place to cirri, these last afterwards dispersing in like manner, until at length a cloudless state of the atmosphere ensues. It is, then, the further passage onwards to S.E. (the hottest wind in summer) of this *lower* current, the higher one still coming from the same quarter as before, which causes the haze, and which appears due to the mixing of two airs of very different temperatures, both however being in a very dry state. A condensation of vapour immediately takes place, but, from the high temperature of the south-easterly current, the particles of the mist, which are probably extremely small, remain suspended, and for a time are equally diffused. As the temperature continues to increase, these particles, perhaps from some change in their electric state, instead of forming cirro-stratus, and passing into rain, become aggregated into cumuli, the form of cloud so peculiarly characteristic of fine summer weather.

(60.) The great heat on the 1st of August, 1846, when the thermometer rose to 88°, was preceded by the same haze, attendant upon the wind, which had been easterly in the morning, shifting to S.E. Here too it was only the lower current, but the barometer was not higher than 29·777 at 10 A.M., and there was reason to believe that, though earlier in the day (as well as for three days previous, on each of which the temperature was above 80°, and the barometer above 30 in.) the upper one was from the N., much of this last veered to the S.W. in P.M., as heavy thunder-storms came up in the evening from this quarter; the haze, in this instance, instead of collecting into cumuli, forming first cirrus clouds, then cirrostratus, then nimbus.

(61.) The kind of haze to which I have been alluding, always occurs under circumstances more or less similar to the above. It seems to be nearly allied to, if not identical with, what has been called by meteorologists *dry fog*, to which I shall have occasion to advert further in another part of this work*.

* In the following instance, extracted from *Laing's Tour in Sweden* (Lond. 1839), p. 162, the phenomenon observed seems to have been nearly allied to the haze I have described above :—
" During the whole evening we had the weather excessively hot,

(62.) *Extreme Seasons.*—The same conditions requisite for producing extremely high and low temperatures, or at least some of them, when operating for a longer time, are probably influential to a certain extent in

although there was a kind of haze over the sky which obscured the sun so much that I could not light my pipe with a good burning-glass. I supposed some large tract of forest must be on fire in the interior, as it was not like fog or mist from evaporation; but the sailors call it *sun-smoke*, and distinguish it from the fogs that carry damp or rain."

Dr. Hooker also mentions this phenomenon in his *Himalayan Journals.* Travelling from Dorjiling to Titalya, he says,—"In the afternoon I rode on leisurely to Titalya,—along the banks of the Mahanuddy, the atmosphere being so densely hazy, that objects a few miles off were invisible, and the sun quite concealed, though its light was so powerful that no part of the sky could be steadily gazed upon. This state of the air is very curious, and has met with various attempts at explanation, all unsatisfactory to me: it accompanies great heat, dryness and elasticity of the suspended vapours, and is not affected by wind." —Vol. i. p. 375.

Mr. Toplis, in his *Observations on the Weather*, thinks the haziness of the air in very hot weather is due in part to refraction, which he explains as follows:—"The misty appearance of the atmosphere during very hot dry weather, will be increased by the mixing of airs of different densities, owing to the great variations of the temperature of the surface of the arid soil from the cloudless days and nights. When the sun rises, the ground is rendered very cold by radiation during the night, and the air above it, to a certain height, has a much reduced temperature. As the earth becomes heated, the air over it expands upwards, and, by mingling with that of a greater density, from the different refracting powers of the two uniting gases, prevents the direct passage of the sun's rays, and causes an opake appearance. This is similar to what takes place in a mixture of alcohol and water before they are completely united."—P. 22.

causing *seasons* of an extreme temperature. It is difficult, however, in all cases, to trace any decided connexion between these conditions and seasons of an unusual character. During the series of years over which my observations extend, I find six summers having a mean temperature *above* the average of the whole series, nine having one *below* it, and four having one *about* the average or within half a degree. The summers with the mean temperature in excess are those of 1831, 1834, 1835, 1838, 1842, and 1846. Only four of these, however, 1834, 1835, 1842, and 1846, had the temperature sufficiently in excess to deserve to be called particularly hot summers.

The summer of 1834 had a mean temperature of 62°·9, being about 2 degrees above the average, as shown in the Table previously given (34). It was preceded by a very mild winter, though the early part of the Spring was cold and backward.

The summer of 1835 was about the same temperature as that of 1834, and was likewise preceded by a very mild winter, though the spring was rather cold and ungenial.

The summer of 1842 was particularly hot in the month of August. The mean temperature of the whole season was 63°·2, being 2°·33 above the average. Preceded by a winter of moderate character, the early part mild; spring seasonable, with a dry April and May.

The summer of 1846 was extremely fine and hot throughout. The mean temperature was 65°·88, or at least 5 degrees above the average. Preceded by a *very mild* winter, and a spring likewise unusually mild and forward. Extreme drought, and almost cloudless skies, for a period of nearly five weeks, from May 21 to June 23.

(63.) It will be seen that these hot summers were all preceded by more or less mild winters. This was particularly the case with those of 1834 and 1846, the latter the hottest in the whole series. I find also, on looking to Mr. Belville's Table of the Temperatures of the seasons at Greenwich for a period of thirty-five years, that the same was the case with some of the hot summers that occurred previously to the commencement of my own observations. Thus the summers of 1822 and 1826, which had respectively a mean temperature of $63°\cdot10$ and $65°\cdot92$, were both preceded by mild winters. This, however, was not so much the case with the summer of 1818, the mean temperature of which was $65°\cdot6$, and which forms, therefore, rather an exception to the supposed rule.

(64.) The coldest summers in the series were those of 1833, 1839, 1841, and 1843, all which had a mean temperature below $60°$, that of 1841 being little more than $58°$; but I can trace no connexion between these cold summers and the character of the preceding winter and spring. The summer of 1833 was remarkable, as having on the whole a rather low temperature, notwithstanding the weather was generally fine and dry: May was the hottest month in that year. Cold summers are usually very wet ones: this was especially the case with those of 1841 and 1843.

(65.) It is an error to suppose that particularly hot summers are productive of the best crops, either in the garden or the field. So far from it, such seasons being generally characterized by long droughts, vegetation is stunted in growth, as well as liable to be burnt up by the excessive heat. Probably summers at or a little above the mean temperature are the most favourable

both to the farmer and the gardener, and in the end
yield the richest produce; and this seems more especially
the case when the spring has been rather backward, so
as to secure the opening buds from the effects of
late frosts. If the summers are much below the mean
temperature, the harvests are late and uncertain, the
crops liable to be mildewed, and fruits do not ripen well,
and never acquire their full flavour.

(66.) The summer in my series which approaches
nearest to the mean temperature of the whole is that of
1847; and it is singular that each season in that year,
commencing with spring, was nearly at a mean, so that
it may be taken as a *type*, to which all other years,
meteorologically considered, have a tendency to conform
in that part of Cambridgeshire in which the observations
were made. This will appear on the following comparison :—

	Spring.	Summer.	Autumn.	Winter.
Mean temp. on an ave- rage of 19 years .... }	47°·18	60°·87	49°·93	38°·09
Mean temp. of 1847 ..	47°·70	60°·88	49°·65	39°·13

It is interesting, therefore, to inquire the more closely
into its character; and the first noticeable point, in
reference to its productiveness, is, that the previous
summer had been a particularly hot one, which may
have had an effect on fruit-trees (according to the theory
of some horticulturists), in well-maturing the shoots of
the year which become the bearing wood of the next.
It is also to be remarked, that the winter immediately
preceding had been very severe, and the cold weather
protracted to a late period: this rendered vegetation
very backward, and even by the end of April the greater
part of the forest trees were still naked. May, very

different from April, was a fine spring month, with seasonable rains, and not a single frost; vegetation, consequently, received no check after it had once made a start, and the garden crops, as well as the foliage of the trees, appeared in great luxuriance at the beginning of the month: the wall-fruit set well, and in great abundance. The temperature of May was rather above the mean, that of June rather below it; but the weather during the latter month, though changeable, was never unseasonably cold, and the rain, which was about the mean quantity, was decidedly favourable both to the garden and the field. July was, on the whole, a fine summer month, the temperature a little above the mean, the quantity of rain rather below it. August was nearly of the same character. Wheat-cutting commenced on the 30th of July *; the crops were heavy, and every-where well got in, while in the garden there was an abundance of most kinds of fruits.

(67.) With regard to the winters during the series of years from 1831–49, I find nine with a mean temperature above the average, seven below it, and three at or about the average temperature; and it is interesting to notice how exactly the *mild winters* balance in number the *cold summers* already alluded to, so as to preserve the mean yearly temperature, in the long run, within fixed limits.

(68.) Of these mild winters five had a mean temperature above 40°, being those of 1833–34, 1834–35, 1842–43, 1845–46, 1848–49. The mildest of all was that of 1833–34, the mean temperature of which was 42°·93, or nearly

* This is exactly the *mean* day on which I have ascertained wheat-cutting to commence, on an average of twelve years, in the parish of Swaffham Bulbeck.

5° above the average. Most of them had been preceded
either by a fine though not always hot summer, or a very
fine autumn. The quantity of rain in particular had
been generally more or less below the mean fall during
each of those seasons; and this seems in accordance
with the converse observation, which has been sometimes
made, that severe winters generally occur after cold
wet summers and autumns. Thus White, in his *Natural
History of Selborne*, remarks that "there is great reason
to believe that intense frosts seldom take place till the
earth is perfectly glutted and chilled with water; and
hence dry autumns are seldom followed by rigorous
winters*." It also falls in with one of Howard's conditions for causing high and low temperatures respectively; viz. *a dry and warm state of the soil*, to some
considerable depth, in the former instance, and *a cold
state of the soil* (the usual result, he adds, of previous
cloudy wet weather) in the latter. It is obvious that
the cloudy skies, which always accompany wet seasons,
must have the effect of checking the power of the sun's
rays, and causing the earth's surface to be much less
heated during summer and autumn, if these seasons
be wet, than when the skies are bright. Of course,
under these circumstances, when the cold of winter sets
in, it leads to a much further reduction of temperature
than would have otherwise occurred. After a very fine
season, the contrary takes place, as the more heated any
body is before it is exposed to cooling influences, the
longer it will be before its temperature falls to a given
point. We might expect, therefore, that the winter

* Letter LXII. to Daines Barrington. White adds in a note,
that "the terrible long frost in 1739–40 set in after a rainy
season, and when the springs were very high."

after such a season would be mild, or, if severe, that it would not set in, as often is the case, till very late. Yet that no *certain* connexion exists between dry summers and autumns and mild winters, appears from the instance of the winter of 1848–49, the mean temperature of which was 41°, notwithstanding the rain had been greatly in excess throughout the whole preceding summer and autumn. This shows how careful we must be in applying any general rule about the weather to particular cases, and how complex probably are all the conditions which must be taken into account to enable us to anticipate with any degree of confidence the character of the coming season.

(69.) It is a curious circumstance, and a striking proof of the extreme uncertainty of an English climate, that, during seasons of an unusual character, we sometimes have the same temperatures occurring for one or more days at the two opposite periods of winter and summer. I have known the thermometer on Christmas-day higher than it had been on the Midsummer-day preceding. On the two last days in June 1839, the temperature never rose higher than 54° (on the 29th) and 53° (on the 30th), while on the 23rd of December following it got up to $55\frac{1}{2}$°. On three consecutive days in June 1841 (the 7th, 8th and 9th), the maximum temperature did not exceed 52°, 54° and 55° respectively, while on each of the two last days in December 1842 it was also 55°. On the 25th of June, 1835, the maximum temperature was actually as low as $51\frac{1}{2}$°, while in December 1848, for seven days in succession, it was never lower than 53°, several times as high as 55°, and on one occasion (the 8th) as high as 58°. This last was the highest temperature I ever registered in the month of December*. During most of these very mild days the weather was extremely fine, the thrush was heard in full song, peacock butterflies were on wing, and, as far as the appearances of nature went, it was difficult to persuade oneself it was not spring instead of winter.

(70.) Of the cold winters during the above series of years, only four were particularly severe, viz. those of 1837–38, 1840–41, 1844–45, and 1846–47. None of these winters had a mean temperature exceeding $34\frac{1}{2}°$. The severest of all was that of 1840–41, the mean temperature of which was only $32°·73$, being more than $5°$ below the average mean, and only a little above the freezing-point. I find no other winter with the temperature lower than this, or indeed so low, during the whole series of years in Mr. Belville's Table, reaching from 1815 to 1849. The one which approaches it most nearly is that of 1829–30, the mean temperature of which is set at $32°·87$. These are probably two of the severest winters that have occurred in this country since that of 1813–14, which was also a very severe one†.

(71.) There was nothing remarkable either in the summer or autumn previous to the hard winter of 1840–41. The weather throughout both those seasons was rather cool and changeable, but the rain was not in excess, and the quantity that fell during the summer especially was more than an inch and a half below the average.

(72.) I find the difference in the mean temperature

* At Upper Swainswick, near Bath, I saw the thermometer at 60°, one day in December 1856.

† The coldest winter on record in this country was that of 1794, the mean temperature of which, at Greenwich, according to Mr. Glaisher, was only 31°·6.—*Phil. Trans.* 1850, p. 594.

of the *mildest and the severest winter* in the series of years over which my observations extend, to be rather more than 10 degrees. The difference between the mean temperature of the *hottest and the coldest summer* in the same series of years is not quite 8°. It would seem by this that the winters in our climate have a wider range of variation, or differ more from one another, than the summers.

(73.) Mr. Graham Hutchinson, in a communication made to the British Association at Glasgow in 1840, has suggested what he conceives to be "a method of prognosticating the probable mean temperature of the several winter months from that of corresponding months in the preceding summer*." But neither in the above instance, nor in that of any other particular year, can I detect any such connexion between the two as he speaks of. His method is founded upon the fact of "the slowness with which the increased temperature of summer penetrates the surface of the ground," in consequence of which, as the author supposes, "the last portion absorbed during the summer half of the year, and which descends to the least depth below the surface, should be the first portion given off during the winter half; and in like manner, the first portion absorbed during the summer half, and which must descend to a greater depth below the surface than any other portion, should be the last to be given off during the winter half." On this principle, each month of the summer half of the year, during which an absorption of heat takes place, should have a corresponding month in the winter half, during which the same heat should be again given out; and the actual temperatures of these two corresponding

* *Report of Brit. Assoc.* (Trans. of the Sections, p. 41.)

months in any year should bear a similar ratio to their
respective mean temperatures. Thus, if a given summer
month be warmer than the average, its corresponding
winter month should be so likewise; if colder, the corresponding month should also be colder. These corresponding months are stated by the author as follows :—

August has October following
July   „   November „   for its correspond-
June   „   December „   ing month of tem-
May   „   January   „   perature.
April   „   February „

But however correct this may be, so far as the temperature of the winter months is dependent upon the one
cause above referred to, it is obvious that there are so
many other causes which have more or less influence in
determining this temperature, that no safe prognostications in any particular year could be grounded upon
this theory. Of this the author, indeed, is himself well
aware; and he allows that these other causes, such as
the proportion of northerly and southerly winds, and
the amount of rain that falls during the same winter
months, varying greatly in different years, and which
are quite independent of the cause under consideration,
may either cooperate with this last, or act in opposition
to it. If it is possible, therefore, to detect any real connexion in the way supposed between the temperatures
of the respective summer months and the corresponding
winter ones, it can only be by comparing their respective
temperatures in a succession of years, by which the
effects of these disturbing causes are in some measure
neutralized. Mr. Hutchinson says he has done this,
and from the results obtained thinks "that in Scotland
deviations in the mean temperature of the summer

months have a visible influence in producing like deviations in their corresponding months of temperature in
the subsequent winter half of the year." I cannot say,
however, that in Cambridgeshire I see any more decided
proof of this influence in the case of a series of years
than in that of particular years taken singly. On comparing the mean temperatures of the corresponding
months, as above given, over a series of nineteen years,
not only are the deviations from the mean very dissimilar
in most instances, but they are just as often on *different
sides* of the mean as on the *same side*; that is to say,
instances in which the two corresponding months are
found to be both warmer or both colder than usual, are
not more numerous than those in which one is warmer
and the other colder, or *vice versâ*.

(74.) *Temperature differently influenced by clouds and
rain, according to the season.*—Almost every one must
have noticed the different effect of bright and cloudy
weather upon the temperature in summer and winter
respectively. On a fine summer's morning, if the sky
is clear, and so continue all day, the temperature increases as the sun approaches the meridian, and for two
hours or more after it has crossed it; and very great
heats never occur except under such circumstances. If,
on the contrary, at or a little before noon the sky should
become generally overcast, the temperature remains
nearly stationary for some hours, or may even fall below
what it was earlier in the day; and cold days in summer
are always attended by thick or clouded skies. In
winter these effects are reversed: the mildest weather
at this season is mostly accompanied by clouded skies,
and cold frosty weather by clear skies. The reason
seems to be this:—During summer, the earth receives

more heat from the sun than it gives out by radiation; any great excesses of heat, therefore, are dependent mainly upon the intensity of the sun's rays, though no doubt in connexion with other causes. In winter, on the contrary, the earth gives out, during the greater part of the twenty-four hours, more heat than it receives; and while this process is favoured by clear skies and a serene atmosphere, it is checked by clouds, which intercept the heat that would otherwise radiate into space, and reflect it back to the earth. Add to this, that mild weather in winter is due always to south-westerly winds, which are more or less charged with vapour, and occasion the cloudiness required for cutting off the escape of terrestrial heat.

(75.) A similar difference is observable in the effect of *rain* upon the temperature, according to the season. In summer a hard shower causes a reduction of the temperature: the effect of a thunder-storm, when attended as it usually is by heavy rain, in cooling the air is matter of common remark. In winter, on the contrary, rain causes an elevation of the temperature. This difference arises from the circumstance of the rain in summer falling from more elevated regions than in winter, and which are consequently much colder than those below. Hence the rain having a great capacity for heat, readily absorbs it from the warmer strata of air through which it passes in its descent to the earth. The cold thus occasioned is further increased by the evaporation, which takes place afterwards from the earth and all bodies which the rain has wetted*.

(76.) In winter the rain falls from a lower elevation, and is moreover generally caused by S.W. winds bring-

* See *Kämtz's Meteorology* (by Walker), p. 155.

ing up vapour of a higher temperature than the earth,
and that portion of the atmosphere in contact with it.
By the condensation of this vapour latent heat is given
out, and the temperature of the earth is raised instead
of lowered.   The evaporation also at this season is very
slight, if not reduced to its lowest point.

(77.) *Solar Radiation.*—It was stated, when speaking
of the cautions to be attended to in fixing a thermometer, that the instrument, if care be not taken, would
be liable to be affected by radiation; that is to say,
instead of correctly indicating the temperature of the
air, it would be raised above, or fall below that temperature, according as it received the rays of heat ema-
nating from other bodies in the neighbourhood warmer
than itself, or gave out heat to objects colder than itself.
The two cases in which this deviation from the true
temperature is most observable, are those in which, on
the one hand, the instrument is exposed by day to the
direct rays of the sun itself, or on the other, allowed to
radiate freely upwards to the sky during the night, and
in these cases it is desirable to ascertain the amount of
the effect produced.   To obtain the greatest effect, it is
necessary that the sky be perfectly clear, the air still,
and the place of observation as open as possible.   The
radiating thermometer may be placed on the ground,
or, what is better perhaps for measuring solar radiation,
raised about an inch above it; and another thermometer
for measuring the temperature of the air, with which it
is to be compared, suspended near, in a sheltered situation, at an elevation of 5 or 6 feet.   The bulb of
the radiating thermometer should project beyond the
frame to which it is attached, and be either blackened,
or covered with black wool, to render it more absorbent

D

of the sun's heat. Observations of this kind are not without value, if properly conducted. It has been shown, indeed, that the difference between two such thermometers is not a correct measure of the force of solar radiation, as has sometimes been supposed, other causes besides the sun's rays combining to affect the temperature of the instrument exposed in the way just described. But to those who do not possess Sir J. Herschel's *Actinometer**, more especially if their chief object is to get a general idea of the conditions of heat to which vegetation is exposed under a bright sun, rather than to determine the laws of heat itself, it may be sufficient to have recourse to the above arrangement.

(78.) Solar radiation begins to take effect soon after sunrise, and, if the sky keep clear, gradually increases as the sun approaches the meridian. Soon after that, it declines, until towards sunset; when, the earth no longer receiving any supply of heat from the sun, it ceases altogether. It is then succeeded by *terrestrial* radiation. The direct influence of the sun's rays arising in this way is of the greatest importance in ripening seeds and fruits. If the sky be constantly clouded at the season of the year at which they attain maturity, whatever may be the mean temperature of the air, seeds ripen imperfectly or not at all, while fruits fail in acquiring their full flavour. This is well known to gardeners and agriculturists, though they seldom give the subject much investigation. It would be very interesting, however, if such persons were more frequently to keep calendars of the several periods at which fruits and seeds ripen in different seasons, at the same time

* For an "Explanation of the Principle and Construction of the Actinometer," see *Rep. Brit. Assoc.* 1833, p. 378.

marking those in which they were most abundant, or in which the ripening process was most successful*. Thus the year 1845 was distinguished by its cool summer, with the sky generally overcast : the month of June was bright, but the whole of July, and the greater part of August, were remarkably cloudy. This state of weather had a visible effect upon the fruits of that season. Those which ordinarily ripen during the month of June, or the beginning of July, such as strawberries, raspberries, gooseberries, and currants, were all in great plenty and luxuriance, though gooseberries, which are usually a little later than the others above mentioned, came too much under the influence of the dull skies of July to be ready for the table before the 23rd of that month. But the wall-fruit, which never ripens till a much later period of the summer, was very late indeed that year, and of indifferent flavour when gathered. The same season was remarkable for the latest harvest I ever noticed. Wheat, which on a mean of twelve years is cut in Cambridgeshire on the 30th of July, was not cut that year till the 18th of August. It had *flowered* on the 26th of June, which is not above three days after the mean ; the lateness of the harvest, therefore, was due entirely to the long run of cloudy weather that set in immediately after the flowering had taken place. It would be of much interest to determine, not merely in the case of corn, but in that of fruits and seeds generally, what the difference of effect may be, according as solar radiation exerts its full power or not, at the

* Such calendars should be kept in connexion with a register of the weather, noting especially, if not the actual heat of the sun's rays each day, at least the comparative amount of sunshine and cloud.

two periods of flowering and ripening respectively. In order to have fruit and seed at all, there must be first a due fertilization of the germen; to use the gardener's expression, "the fruit must set;" and it is quite as important to have the weather favourable when this process is going on, as it is afterwards, when the fruit, which has well set, is beginning to ripen. I regret I have no observations of my own to offer on this important question, but the result one might perhaps be led to expect is this, that the abundance of the crop would depend upon the character of the weather at the time the fruit was setting, while the size and luxuriance of the grain or fruit, taken individually, would be more due to bright skies and a high temperature during the interval between the setting and the ripening.

(79.) There are two series of observations connected with solar radiation which it is desirable should be made in as many localities as possible. One has reference to the relative intensity of the sun's rays during each successive month in the year; the other to the progress of the same from hour to hour during the day. The variations under the first head have a direct bearing on those questions, connected with the ripening of fruits and seeds, just alluded to, while those under the second, assist us in determining, what is also of great importance, the maximum degree of heat to which a plant is subjected, when exposed to the full influence of a bright mid-day sun.

There are, however, but few observations of either kind, carried on continuously for any length of time, at least in this country, on record. Those given by Daniell, in his "Meteorological Essays*," are almost

* 1st Edit. p. 210, &c.

the only ones I am acquainted with, besides a very few
imperfect ones of my own. Both his and mine are
given side by side in the following Table, which shows
the mean maximum temperature of the air, with the
mean maximum, as well as the maximum, excess of the
thermometer in the sun, for every month in the year.
It must be borne in mind that the entries under the
second head are the *differences* between the indications
of the radiating thermometer and the temperature of
the air. Daniell's observations are the result of something less than two years' registering. Mine are the
result of only one, reaching from November 1848 to
October 1849. They were made with one of Newman's
Radiating Thermometers placed in an open spot upon
a lawn that was kept mown, the bulb of the instrument,
which was blackened and entirely exposed, being raised
about an inch above the grass.

TABLE V.—*Showing the mean maximum temperature of
the air, with the mean and maximum power of the sun,
for every month of the year.*

Month.	Mean maxim. of the Air.		Mean maxim. excess of Radiating Therm.		Maxim. excess of Radiating Therm.	
	Daniell.	Self.	Daniell.	Self.	Daniell.	Self.
January	39·6	40·1	4·4	9	12	24
February	42·4	47·5	10·1	13·5	36	30
March	50·1	49·3	16	18·8	49	34
April	57·7	54·1	28·1	24·8	47	51
May	62·9	62·2	30·5	21·1	57	31
June	69·4	67·3	39·9	26·3	65	36
July	69·2	70·5	25·8	24·2	55	38
August	70·1	70·1	33·1	22·6	59	32
September	65·6	65·3	32·7	24·5	54	30
October	55·7	55·8	27·5	17·5	43	31
November	47·5	47·2	6·7	14·7	24	25
December	43·2	46·4	5·4	8·2	12	19

(80.) It appears from both Daniell's observations and my own in the above Table, that the mean greatest intensity of the sun's rays occurs in June, the same being the month of the sun's greatest declination. The coincidence of the two is what, as Daniell observes, might be anticipated. His observations, however, show a much more regular progression than mine in the advance of the mean towards the maximum, while the maximum itself, especially during the summer months, is considerably higher. The former circumstance may be due in part to my observations having been more imperfectly conducted, and for a shorter time than his. The latter difference may arise from the bulb of his thermometer having been covered with black wool, while mine, though blackened, was naked; wool being a more powerful radiator than glass. It is evident, however, that the observations of a single year must be always liable to present great irregularities. As the heat of the sun's rays depends so much upon the state of the sky and clearness of the atmosphere, it must not unfrequently happen that in a given year the maximum, and even the mean maximum, may be displaced by a cloudy season, and found in a different month from that to which the observations of a long series of years would otherwise refer it. Thus, in the year in which I made my observations, the maximum intensity occurred as early as April: this was due, not to any particularly hot weather, but rather, on the contrary, to sharp frosts, with their usual accompaniment of a clear sky, occurring late in the spring, when the sun's meridian altitude, and by consequence the intensity of his rays, were very considerable. The intensity being measured by the excess of the radiating thermometer above the temperature of the air, the greatness of the effect may arise quite as much from the latter being low as the former being high.   In fact, it will be often found that when the absolute height of the radiating thermometer is at its maximum, the temperature of the air is at its maximum also, causing the difference between the two to be comparatively small.

(81.) Almost all the days of the greatest intensity of solar radiation occur with north, north-easterly, or north-westerly winds; very seldom with south-westerly. This is easily explained; the former winds being drier than the latter, with less vapour in the air to intercept the sun's rays.   It is this circumstance, indeed, which renders solar radiation so powerful during clear frost, as in the instance above mentioned, northerly winds generally prevailing at such times.   The difference in the effect, according as the wind blows from N.E. or S.W., is strikingly seen in the following instances of two days in the same month of January, and in the same year as that in which the observations given in the above Table were made.

	Maxim. temp. of the Air.	Maxim. Rad. Therm.	Differ- ence.	Wind.	Weather.
Jan. 6......	25°	45°	20°	N.E.	Frost very severe the whole day, with bright cloudless sky.
Jan. 19 ...	52	59	7	S.W.	Extremely fine and mild: sky clear.

Both these days were equally fine, and the sky equally free from clouds.   Nevertheless it will be observed, that, though the radiating thermometer rose 14 degrees higher on the 19th, when the wind was S.W., than on the 6th, when the wind was N.E., the *difference* between it and the temperature of the air was nearly three times as great on the latter of these days as on the former. On the 6th it was more than *double* the mean maximum excess of the radiating thermometer for that month, as shown in the Table, while on the 19th it was 2 degrees *less* than the mean.

(82.) This difference, depending upon the direction of the wind, is not confined to winter, or times of frost. The same results, though less striking, were obtained on two consecutive days, during some very hot weather in the month of April, in the year 1848, as under:—

	Maxim. temp. of Air.	Maxim. Rad. Therm.	Differ- ence.	Wind.	Weather.
April 1 ...	67°	93°	26°	N.W.	Extremely fine and hot: sky nearly cloudless the whole day.
April 2 ...	71·5	91	19·5	S¹ʸ	As yesterday.

Here it will be seen, that, on the 2nd, with the southerly wind, though the temperature of the air was 4½ degrees higher than on the 1st, the radiating thermometer was 2 degrees lower, and the difference, indicating the intensity of the sun's rays, 6½ degrees less.

(83.) The clearness of the atmosphere during fine weather with northerly winds, is sometimes even a compensation for a light breeze, which, *cæteris paribus*, has always the effect of lessening the intensity of the sun's rays. Thus, on January 26, during a fine bright day, but attended by a cool-blowing wind from N.W., the excess of the radiating thermometer was 24 degrees, while on the 20th of the same month, also extremely fine and quite calm, but with the wind S.W., the excess had not been higher than 15 degrees.

(84.) I have made but few observations on the subject of the progression of solar radiation during the day, and the exact time of the maximum. Such, however, as I have made may be worth recording. For conducting experiments of this kind, it is necessary to select a bright cloudless day free from wind. Daniell has given the result of some made by himself in the month of June, by which it appears that the maximum force of the sun's rays, as he terms it, or the greatest difference between the radiating thermometer and the temperature of the air, occurs, on an average of five days' observations, at 1^h 30^m P.M. This agrees with the results obtained by myself in the same month, but not with all those obtained in other months. Whether season exercises any influence in this matter, can only be determined by repeated observations made in each month of the year, or at least at each of the four quarters. In a set of observations made by myself on the 12th of February, 1849, and given on the next page, when the weather and state of the sky were extremely favourable throughout the day, the maximum difference between the two thermometers occurred as early as 11^h 30^m A.M., continuing unabated till noon. On a previous occasion in the same year, in the month of January, during frost, with a bright cloudless sky and the wind N.W., the conditions remaining unchanged during the day, the maximum difference, equalling 21 degrees, occurred at 30^m after noon. The time of the maximum difference may or may not coincide with the time of the greatest absolute height of the radiating thermometer. Both are subject to a little variation, though occurring, on an average, about one o'clock or half-past. Sometimes the maximum height of the radiating thermo-

meter is earlier than it would otherwise be, owing to
an increase of vapour in the air, raised from the ground
by the sun's heat, as the latter gets higher in the
heavens*.   The temperature, also, does not always rise
by regular advances, but is subject to slight fluctuations,
running back occasionally a few degrees, in consequence,
apparently, of light currents in the air at intervals, as
the latter becomes more rarefied.

TABLE VI.—*Progress of solar radiation from hour to
hour on the 12th of February,* 1849.

Sky bright and cloudless the whole day.   Wind N.W., quite
calm.   Barom. at 10 A.M. 30·795.

Thermometer.			
Time.	Exposed.	In shade.	Difference.
A.M. 9	44	36·5	7·5
10	55	39·5	15·5
10½	63·5	40	23·5
11	68	41	27
11½	69	41·5	27·5
12	70	42·5	27·5
P.M. 1	71	45	26
1½	71	46	25
2	63	46·5	16·5
3	56·5	47	9·5
4	45	45·5	·5
4½	39·5	44	4·5
5	34	41	7
Sun just setting, and stratus forming.			
6	31	37	6
7	29·5	35	5·5

* A particular instance of this is noticed by Dr. Hooker, in
his *Himalayan Journals,* where he speaks of the black-bulb
thermometer as rising in the sun, among the hills of Behar, to
130°, as early as 9½ A.M.   And he adds, " the morning observation before 10 or 11 A.M. always gives a higher result than at

The above observations were continued till after the time when solar radiation gives place to terrestrial; and it appears that at this period of the year the temperatures in the sun and shade are reduced nearly to an equality by 4 P.M. After that hour the radiating thermometer begins to fall below the other, the greatest difference occurring about sunset.

TABLE VII.—*Progress of solar radiation on the* 30*th of June,* 1832.

Bright and nearly cloudless sky. Air quite calm. Wind S. Barom. at 10 A.M. 30·401.

Thermometer.			
Time.	Exposed.	In shade.	Difference.
P.M. 0$\frac{1}{2}$	119	70	49
1	119	70	49
1$\frac{1}{2}$	127·5	72	55·5
2	120	72	48
2$\frac{1}{2}$	113	72	41
3	115	72	43

In this instance the maximum temperature in the sun and shade, as well as the maximum excess of the radiating thermometer above the other, were all coincident at 1^h 30^m P.M., though the temperature of the air continued the same till 3 P.M.

Similar observations were made in this month, on the 26th day, in a subsequent year, with one difference in the attending circumstances, which is worth noting. The weather as before was extremely fine, with a bright cloudless sky from morning to night, the wind S.E., and quite calm; but though the maximum height of noon, though the sun's declination is so considerably less,—an effect no doubt due to the vapours raised by the sun."—Vol. i. p. 15.

the radiating thermometer, as well as its maximum excess above the temperature of the air, both occurred at 1 P.M.—half an hour earlier than in the above instance,—the temperature of the air itself, which had been 78° at the hour just mentioned, by an unusual accident kept increasing till 5ʰ 30ᵐ P.M., when it attained to 80°·5. During these 4½ hours the radiating thermometer slightly fluctuated, rising and falling by turns. It never fell more than 4 degrees, and at 3 P.M. was within half a degree of what it had been at 1ʰ 30ᵐ

TABLE VIII.—*Progress of solar radiation on the 25th of September*, 1832.

Sky perfectly cloudless.   Air very still.   Wind S.   Barom. at 10 A.M. 30·398.

Thermometer.			
Time.	Exposed.	In shade.	Difference.
A.M. 11ʰ	112·5	69	43·5
11½	115	71	44
12	119·5	71·5	48
P.M. 0½	118·5	72·5	46
1	120	73	47
1½	117	73·5	43·5
2	117	74	43
2½	115	74·5	40·5
3	108	74·5	33·5

Here the maximum excess of the radiating thermometer occurred exactly at noon, though its greatest absolute height was not till an hour afterwards.

(85.) *Terrestrial Radiation.*—This is due to the earth giving out the heat at night which it has received from the sun during the day. It commences in general about an hour before sunset, and keeps increasing as the sun approaches the horizon. The maximum, under a clear sky, appears to be attained just at sunset itself, or shortly after. Both the radiating thermometer, and that which indicates the temperature of the air, may fall lower during the night, and if the state of the atmosphere remain the same, they will continue to fall till towards morning. But the difference between these two thermometers, by which the terrestrial radiation is measured, will be found generally to diminish soon after sunset, and to get less, until it reaches a certain point at which it becomes stationary, or nearly so. Terrestrial radiation, however, is very considerable again at sunrise; and if the early part of the night have not been favourable for it, it is often greatest at that time, and equal to what ordinarily occurs at sunset. As far as my observation goes, the mean maximum amount of the difference between the two thermometers at sunset, when the state of the sky and weather are such as to allow of radiation having its full effect, is about 10 degrees.

(86.) As the power of *Solar* Radiation is thought by Daniell to follow the course of the sun's declination, the maximum intensity and effect occurring in June, so also it is considered by that observer that *Terrestrial* Radiation has an evident tendency to increase with the heat, the maximum depression of temperature from this cause occurring in June likewise. The mean depression, however, he found greatest in April. My own observations are too scanty and imperfect to throw much light on this point, but such as they are, they are given by the side of Daniell's, as in the former instance, in the following Table. His are deduced from the averages of three years: mine are the results of only one.

TABLE IX.—*Showing the mean minimum temperature of the air, with the mean and maximum amount of terrestrial radiation, for every month in the year.*

Month.	Mean minimum of the Air.		Mean depression from radiation.		Maximum depression from radiation.	
	Daniell.	Self.	Daniell.	Self.	Daniell.	Self.
January	32·6	32	3·5	3·3	10	9
February	33·7	36·7	4·7	4·1	10	13·5
March............	37·7	37	5·5	3·8	10	9
April	42·2	40	6·2	5	14	10
May	45·1	46·7	4·2	4	13	8
June	48·1	48·6	5·2	5·8	17	11
July	52·2	52·2	3·6	5·2	13	11
August	52·9	54	5·2	5·1	12	9
September	50·1	51·4	5·4	3·7	13	11
October	42·1	44·5	4·8	3·5	11	8
November	38·3	36·2	3·6	4·2	10	8
December	35·4	37·8	3·5	3·1	11	6

(87.) Considering that my own results in the above Table are deduced from the observations of a single year, they make a near approximation to those of Daniell's. The approximation, however, is greater in respect of the mean depression than of the maximum. According to both, the *mean* depression is least in winter; though we do not agree in the season in which it is greatest, Daniell making it in the spring, and myself in the summer. The *maximum* depression with me occurred in February, equalling 13½ degrees. The greatest observed by Daniell was 17 degrees, in the month of June. This last is probably as great a depression arising from terrestrial radiation as ordinarily occurs in this country, and is not often exceeded; though it has been supposed by Wells that certain localities might be found in other countries, the circumstances of which, under a perfectly clear sky on a still night, would be favourable for causing a depression of a radiating thermometer to the amount of 30, or even, perhaps, 40 degrees*.

(88.) The necessity of a clear sky as one of the conditions favourable for radiation, whether solar or terrestrial, is remarkably seen by the effect of clouds suddenly supervening to dull the transparency of the atmosphere. By day, the radiating thermometer immediately falls; by night it rises. Almost the thinnest cirrus clouds, at that great elevation at which such clouds usually float, have a sensible influence in this way. They serve in the one case to intercept in part the sun's rays that would otherwise strike the earth; in the other, to reflect back to the earth some of the heat that the earth radiates upwards during the night. If the clouds become very dense, and the sky is entirely overcast, radiation ceases altogether. The point at which the radiating thermometer then stands coincides with the temperature of the surrounding air.

* *Philosophical Essays* (Essay on Dew), pp. 200, 201. The above is probably rather overstated. Persons who wish to see further and more recent researches than those of Wells and Daniell, on the subject of terrestrial radiation, are referred to an elaborate and valuable paper by Mr. Glaisher in the *Phil. Trans.* for 1847, p. 119. In this memoir, the author gives the results of numerous observations made by himself on the amount of heat radiated from the earth at night, and from a great variety of bodies, placed on or near the surface of the earth, and fully exposed. In one instance he observed a difference of 20° between the temperature of the air and that of a thermometer on long grass (p. 145). The body which appeared to radiate most freely was *raw wool*, between the temperature of which, and the temperature of the air, at the height of 8 feet from the ground, there was on one occasion a difference of 28°·5. This was the greatest difference he ever saw (p. 146).

(89.) But there is a phenomenon relating to terrestrial radiation at night, occurring in certain states of weather, which is still more remarkable; and that is, that occasionally the radiating thermometer, instead of merely rising, under the influence of a cloudy sky, to the temperature of the air, will be found *higher* than that temperature. This anomalous circumstance, as at first it appears, seems to be the effect of a sudden increase of temperature in the night beyond what prevailed the day previous; such as not unfrequently takes place in winter, on the change from frost to thaw, as also in spring and autumn, but never that I noticed in summer, the nights being hardly ever hotter than the days at that season. This increase of temperature is due to the wind shifting from some point more or less northerly to W. or S.W., and bringing up, not only currents of warmer air, often at a greater elevation than those nearest the earth, but *clouds charged with warm vapours*, which radiate their heat downwards, and thus cause a rise in the radiating thermometer. The rise of the thermometer which marks the temperature of the air is more gradual, the air being only mediately affected through conduction from the earth, with which it is in contact.

(90.) The following instances in which this occurrence took place are extracted from my Meteorological Register.

(Ex. 1.) 1848, Feb. 5.—Radiating thermometer last night 3 degrees higher than the other*. Mild and generally clouded

---

* By this it is meant that the lowest point to which the radiating thermometer fell was 3 degrees higher than the minimum temperature of the air.

throughout yesterday : temperature of the air at 10 P.M. 48°; at 10 A.M. this morning risen to 53½°.

(Ex. 2.) 1848, Feb. 9.—Radiating thermometer last night 1½ degree higher than the other.   Yesterday mild, and fine till towards evening ; wind then veering to S.W., bringing up clouds and rain.   Temp. of the air at 10 P.M. 46°; at 10 A.M. this morning risen to 49°.

(Ex. 3.) 1848, Oct. 3.—Radiating thermometer last night 4 degrees higher than the other.   Fine and mild yesterday, but clouding over in P.M., with rain at night.   Temp. of the air at 10 P.M. 57°; at 10 A.M. this morning risen to 60½°.

(Ex. 4.) 1849, Jan. 19.—Radiating thermometer last night 1 degree higher than the other.   Very fine and mild both yesterday and today, but the temperature of today rather higher. Therm. at 10 P.M. last night 49°; at 10 A.M. this morning 50°.

(Ex. 5.) 1849, Feb. 2.—Radiating thermometer last night 3 degrees higher than the other.   Fine and frosty yesterday morning, with the wind N.W. : towards evening wind shifting to S.W., sky at the same time becoming thick and misty, followed by rain.   Therm. at 10 P.M. 39°; at 10 this morning risen to 44½°. Much milder all day than yesterday.

(Ex. 6.) 1849, March 1.—Radiating thermometer last night 1 degree higher than the other.   Weather yesterday very stormy, with rain and snow in P.M.   Temp. of the air at 10 P.M. 31¼°; at 10 this morning risen to 38½°.

It will be seen that in all the above cases the weather the next morning was warmer than on the evening previous, the increase of temperature amounting in the last instance to 7 degrees, though this was not the one in which the difference between the two thermometers was greatest*.

(91.) The converse of the above phenomenon, viz. the sinking of the radiating thermometer during day

* Mr. Glaisher, in like manner, found that during cloudy nights a thermometer, with the bulb placed in the focus of a metallic parabolic reflector, fully exposed to the sky, was fre-

below the temperature of the air, is by no means uncommon in wet weather, and is manifestly due to the
cold produced by the evaporation of the rain-drops which
fall upon the bulb of that instrument.

(92.) When we consider the heating effect of the
sun's rays during the day, and the great cold caused by
terrestrial radiation at night, we may conceive the immense range of temperature to which vegetation, in
open places, is occasionally exposed during the twenty-
four hours. This unquestionably often amounts to
more than 100 degrees. The maximum height of the
radiating thermometer recorded by Daniell in his second
Table is 143 degrees. This occurred in June, and is
higher than any I have observed myself. But I have

quently higher than a thermometer 4 feet from the ground, and
protected from the effect of radiation. — *Phil. Trans.* 1847,
p. 193.

There are also some observations made by Dr. Chowne, and
published in *The Athenæum* for 1856 (No. 1512, p. 1276), from
which it appears that he too has noticed the effect of an in-
tensely clouded sky at night, in the winter months, in raising
an exposed thermometer above one sheltered from radiation.
The excess of temperature indicated by the former varied with
him on different occasions from ½ a degree to 2 degrees. These
observations were made in London, and the effect was thought
to be due to the increased heat "generated in the metropolis by
the use of furnaces, by domestic fires, by gas-lights, &c."—this
heat being, in the first instance, radiated upwards to the clouds,
and then reflected back from them to the earth. The above
observations of my own, however, being made in the country,
show that, whatever influence the higher temperature of London
may exercise in this way, the same result sometimes occurs in
cases, in which it can be traced to no other source than to the
clouds themselves being heated above the temperature of the
air nearer the earth.

known the radiating thermometer fall at night in that same month as low as $29\frac{1}{2}°$. Supposing, therefore, which is quite possible, that on any occasion these two extremes were to occur in the same place on the same day of that month, the range in that case would equal $113\frac{1}{2}$ degrees*.

(93.) It is also an important fact that there is scarcely a month in the year in this country, in which frosts, arising from terrestrial radiation, are not liable to occur. I have known frosts in June at Swaffham Bulbeck in two or three instances. On the 29th of that month, in 1837, at sunrise, and for a short time afterwards, the ground was white in places with hoarfrost. Again, on the 13th of June, in 1849 (the instance above alluded to), the radiating thermometer fell $2\frac{1}{2}$ degrees below freezing-point. In the same month, in a former year (the exact date I do not remember), on an occasion when I had no thermometer exposed to mark the exact depression of temperature, the existence of a frost was made known to me by the following remarkable circumstance. It was during the hay season, and the weather had been extremely fine and settled for some time; and the heat in the middle of the day was so great, that a labourer who was employed by me in mowing grass, desisted soon after dinner-time, and retired to rest, with the intention of rising in the middle of the night instead to continue his work. He told me the following day, that, on getting up at 3 in the morning and going into the meadow, he not only found it

---

* Daniell is of opinion that "a plant might be so situated, in the month of June, as to undergo all the changes of heat from 154° to 30°," equalling a range of 124°.—*Meteorolog. Essays,* 1st Edit. p. 278.

white with hoar-frost, but his scythe, which he had left out on the grass the afternoon previous, was firmly frozen to the soil, so as to require some slight effort to detach it. The wind at the time was northerly, and it was that which caused so remarkable a difference in the temperatures of day and night. I have likewise known a frost in August. There was one on the 22nd of that month in 1850, at Ampney in Gloucestershire, where I was staying at the time. I had no thermometer with me to indicate the exact degree of cold that took place, but the ground was still white with frost when I rose at 6 in the morning, and in the garden there were unmistakeable marks of its effects upon some of the more tender plants and vegetables*. Even in July, which at least might be supposed an excepted month, the radiating thermometer once fell at Swaffham Bulbeck to $32\frac{1}{2}°$. This was in 1849, the same year as that in which a frost had occurred in June; and though the temperature in that instance was not quite down to freezing-point, it was so near it, that it is difficult to believe it might not sometimes be slightly further depressed, so as to reach 32° itself, if it did not go below it. Of course, in all the other months of the year, frosts are of frequent occurrence†.

* With respect to frosts in August, the following passage occurs in *Evelyn's Diary*, under the date of August 11, 1695:— "The weather now so cold, that greater frosts were not always seen in the midst of winter; this succeeded much wet, and set harvest extremely back."—Edit. 1850, vol. ii. p. 336.

† There is a notice by Mr. J. F. Miller, of the occurrence of *ice* of the thickness of half-a-crown on the glass of some hot-beds at Whitehaven, Cumberland, on the morning of the 4th of July, 1851 (see *Edinb. New Phil. Journ.* vol. liv. p. 50). In a previous year, 1849, he had noticed a "naked thermometer on

(94.) *Variation of temperature with height.*—Though the temperature diminishes as we ascend in the atmosphere*, so that places situated on lofty hills are, on the whole, colder than those on the plains below, yet there is a well-known exception to this law in the circumstance of the *increase* of temperature *at night* in the different strata of air, as they rise above the surface of the earth, up to a certain elevation. This is due mainly, if not altogether, to terrestrial radiation just spoken of, by which the stratum of air next the earth being first cooled, the cold is propagated gradually upwards to the superior strata in succession. Professor Marcet, who made many observations relating to this phenomenon†, states that this increase of temperature

the grass, placed on raw wool, at or below the freezing-point in every month."—*Id.* vol. xlix. p. 63.

* The *rate* of decrease appears to have been differently estimated by different observers, and to vary in different latitudes. Prof. Forbes says that "there is little doubt that it is not uniform, but slower as we ascend," and that it "is most rapid in spring, and least so in autumn."—*Rep. Brit. Assoc.* 1840, p. 57.

The most recent observations on this subject are those of Mr. Welsh during four balloon ascents, made under the direction of the Kew Observatory Committee of the British Association. The general result arrived at was, that "the temperature of the air decreases uniformly with the height above the earth's surface, until at a certain elevation, varying on different days, the decrease is arrested, and for a space of from 2000 to 3000 feet the temperature remains nearly constant, or even increases by a small amount; the regular diminution being afterwards resumed and generally maintained, at a rate slightly less rapid than in the lower part of the atmosphere, and commencing from a higher temperature than would have existed but for the interruption noticed."—See *Phil. Trans.* 1853, p. 338.

† *Mém. de la Soc. de Physique, &c. de Genève*, tom. viii. (1838):

as we ascend, and which, like terrestrial radiation, is most conspicuous about the time of sunset, "however variable it may be either in regard to its intensity or its limit in reference to elevation, is a constant phenomenon in every state of the sky, except in the case of violent winds." He considers the effect as not ordinarily extending beyond the height of 100 or 110 feet*, even under the most favourable circumstances. The maximum effect is immediately after sunset; and "it is in winter, and particularly when the ground is covered with snow, that the phenomenon presents the most remarkable results." Professor Marcet found on one occasion, during a severe winter at Geneva, a difference of $14°\cdot4$ between two thermometers, one placed 2 feet above the surface of the ground, the other being at the height of 52 feet. "The mean difference, calculating upon twelve observations made during the period of excessive cold, between the temperatures of two strata of air separated by an interval of 50 feet, was $10°$ Fahr." "The comparison between the temperature of the air at *two* feet, and at *five* feet above the surface, perhaps presented still more remarkable results than the preceding, regard being had to their great proximity. The difference, calculating from the mean of nine observations (the surface being then covered with snow), was $4°\cdot2$ in favour of the more elevated station; this difference, on the 4th of January, increased to $7°\cdot2$ Fahr."

(95.) There is a phenomenon, closely allied to the above,

his papers on this subject will likewise be found in the *Edinb. New Phil. Journ.* vol. xxv. p. 353, and vol. xxxii. p. 34.

* Wells, apparently on the authority of Six, sets it at 220 feet; but this estimate is probably too high.—*Essay on Dew,* p. 223.

which will be further treated of in a subsequent part of
this work under the head of Mist,—I allude to the circumstance of the air in valleys and hollows being colder
at night than on more elevated ground adjoining, the
elevation of this last being inconsiderable. I remember
on one occasion, experiencing in a remarkable manner
this difference in the temperature of the air at small
differences of height, at the decline of day. I was
walking in autumn with a friend, about sunset, in the
neighbourhood of Bristol, along a road which was alternately up and down for a considerable distance, owing
to the unevenness of the ground. It presented, in fact,
a succession of low hills and corresponding valleys.
Every time that we descended into one of these hollows we were struck by a sensible chilliness in the
air, which ceased as we rose upon the next emi-
nence, the temperature returning to what it was before.
This continued through the whole series of elevations
and depressions; and my friend, who, as well as myself
at that time, was ignorant of the cause, observed, that
it was like passing alternately from a room in which there
was a fire to one in which there was none, or from the
inside of a house in winter to out-of-doors and back
again. Not, however, having any thermometer with
us, we were unable to note the exact difference in the
temperatures thus experienced.

(96.) *Sensible cold at sunset.*—The late Dr. Heberden
employed the term "sensible cold" to signify "the
degree of cold perceptible to the human body in its
ordinary exposure to the atmosphere," as distinguished
from the temperature of the air itself:—"for," as he
observes, "while the thermometer truly marks the
temperature of the medium in which it is placed, the

sensations of the body depend altogether upon the rapidity with which its own heat is carried off*."

This "sensible cold" is very remarkable about sunset in certain localities, especially in the autumn and winter months, and the more so from its often exceeding what is felt a few hours later in the evening. Much of the cold experienced by persons at this time is due to radiation, and is so far connected with the subject last treated of. Heat, of course, escapes from the human frame the moment it is exposed to a clear sky after sunset, just as it escapes from the radiating thermometer, or from any other bodies similarly circumstanced, and the more freely in consequence of the high temperature which the human body naturally possesses. But we might naturally expect the sensation of cold arising from this cause to be the same equally at all hours of the night, if the state of the weather and atmosphere remained the same; whereas this is not uniformly the case. Neither would it affect a thermometer so placed as to indicate only the temperature of the air; whereas the cold of which I now speak, or the excess of cold at sunset beyond what occurs at a later hour, does show its influence on such a thermometer. There must, therefore, be some other circumstance besides radiation, necessary to be considered, in order to account for this excess. The fact itself I first noticed at a period of my life when I was in the habit of going out in the autumn, just about the hour of sunset, to dine with some relations who lived half a mile off. I usually walked to their house, and I was occasionally much struck with the sensation of a peculiar chill, on first leaving my own house for the open air, beyond

* *Phil. Trans.* 1826, p. 70.

what I experienced on my return at night. At first I naturally expected that if the cold was such when the sun was only just setting, it would be still greater as the evening advanced; but, on looking at the thermometer immediately on getting home, and comparing it with what it was when I set out, I found, to my surprise, that it was actually higher between nine and ten o'clock, than it had been between five and six. It is necessary to remark that this circumstance was only noticed in certain states of weather. The sky at such times was always clear, the wind generally northerly, and either there had been rain in the earlier part of the day, or there was more or less of *stratus* on the adjoining meadows just at sunset. The following instances in which it occurred, are a few selected from my Meteorological Journal, with the exact particulars of the weather on each occasion.

(Ex. 1.) 1838, Oct. 31.—Steady rain, with N. wind, from before sunrise till 3^h P.M. Sky then clearing, and stratus beginning to form soon afterwards. The temperature, which had risen to $44\frac{1}{2}°$ during the day, falling to 35° at sunset, and the air very chilly. Fine and clear throughout the evening, but sensibly warmer three hours after sunset, and the thermometer risen to $38\frac{1}{2}°$.

(Ex. 2.) 1838, Dec. 18.—Fine and settled all day with a high barometer, and the current next the earth S.W. Very cold at sunset, soon after which, the temperature of the air, which had not been higher than 37° any part of the day, fell to 28°. By 10 P.M. it had risen again to 32°.

(Ex. 3.) 1844, Dec. 30.—Thick and clouded in the morning, with the barometer above 30 inches, and the wind N.E. Afternoon clear till sunset, when a dense stratus formed over the low meadows, the thermometer standing at $33\frac{1}{2}°$. Later in the evening, fog generally diffused through the air. Temperature at 10 P.M. risen to 35°.

E

(Ex. 4.) 1845, Dec. 10.—Very fine and clear, with a high barometer, and wind N.W. Maximum temperature during the day
$46\frac{1}{2}°$. Cool and frosty feel in the air about sunset, the thermometer falling to $34\frac{1}{2}°$. At 10 P.M. temperature risen to 37°.

(97.) Of the above instances, that of the 18th of December 1838 is the one in which there was the greatest
rise of the thermometer later in the evening, amounting
to 4 degrees. The wind in this case was S.W., but
this was only the lower current, and no doubt there
was an upper current from some point more or less
northerly, from the great height of the barometer. On
the whole, I am inclined to think (from observations on
the dew-point, in connexion with the temperature of
the air at the time of sunset, which I shall have occasion
to speak of afterwards) that the above phenomenon is
restricted for the most part to low and damp situations,
and that it is mainly due to the *stratus* by which it is
generally accompanied. The fall of the temperature
from terrestrial radiation about sunset is more rapid
than it is later in the evening. The temperature of
the lower strata of the atmosphere, in contact with
the ground, will also be more rapidly depressed at that
time. But no sooner has the temperature of the air
fallen to that of the dew-point (which, if the locality be
damp, or the air moist from previous rain, speedily
ensues), and mist, constituting the stratus, begun to
form, than the depression receives a slight check from
the latent heat which the vapour in the air gives out at
the time of condensation. If the mist rise higher as
the evening advances, the supply of heat from this
source keeps increasing, at the same time that radiation
itself is checked by the interruption caused to the transparency of the air, until the reaction is sufficient to

affect the general temperature as shown by the thermometer.

(98.) Whether the above explanation be correct or
not, and even entirely setting aside all those cases in
which there is any *increase* of temperature afterwards,
there is no question that the *fall* of temperature as the
day declines is more rapid, under a clear sky, about
sunset, than it is later in the evening. And this circumstance, from the chill that it occasions to those who
come suddenly from a warm room into the open air,
renders sunset, perhaps, a more hazardous time for
persons of a weak and delicate constitution to leave the
house, especially if situated low, than almost any other
period of the day or night.

(99.) *Effect of an Eclipse of the Sun on Solar Radiation.*—On the occasion of the great solar eclipse, which
took place on the 15th of May, 1836, when about nine-
tenths of the sun's disc were obscured, I availed myself
of the opportunity of noticing the diminished effect of
the sun's heating power during the occurrence of such
a phenomenon. The state of the weather was extremely
favourable for observations of this nature, being very
settled, as it had been for several days previous, with a
sky nearly free from clouds, the wind N.E., and the
barometer above 30 inches and a half. From the steadiness of the weather, and its being almost exactly the
same on the 14th, the day before the eclipse, as on the
15th, when the eclipse took place, I was enabled further
to institute a close comparison between observations of
the same kind made on each of those days respectively.
The only difference in the weather of the two days was
that, on the 14th, there was a gentle breeze at intervals,
which assisted in cooling the air, while on the 15th

the air was quite calm : this latter in consequence was the hotter of the two days.

(100.) The following Table shows the results of observations on the temperature of the air, and the amount of solar radiation, made every quarter of an hour, from 1ʰ 15ᵐ to 4ʰ P.M., on each of the two days, and the difference between them. The thermometer employed in measuring the temperature of the air was suspended under a tree in the garden, and so placed as to be thoroughly screened from the sky and sun, as well as from all heated objects in the immediate neighbourhood. That used to ascertain the heating power of the sun's rays was placed near the ground, a few inches above a south border, with the bulb, which was not blackened, fully exposed both to the sun and sky. The eclipse commenced at 1ʰ 51ᵐ P.M. The greatest obscuration was at 3ʰ 19ᵐ. The eclipse terminated at 4ʰ 39ᵐ.

Time of observation.	Temp. of air.		Radiating Thermometer.		Difference.	
	May 14.	May 15.	May 14.	May 15.	May 14.	May 15.
1ʰ 15ᵐ	64	67·5	107	109	43	41·5
,, 30ᵐ	64	68	104	111	40	43
,, 45ᵐ	64·5	68·5	103	111	38·5	42·5
2ʰ	64·5	69	103·5	106	39	37
,, 15ᵐ	65	69	102·5	105	37·5	36
,, 30ᵐ	65	68·5	102	100	37	31·5
,, 45ᵐ	65·5	67·5	104	89·5	38·5	22
3ʰ	65·5	67	100	82	34·5	15
,, 15ᵐ	....	65·5	....	71	....	5·5
,, 20ᵐ	65	65	100·5	68	35·5	3
,, 30ᵐ	65	64·5	91	68	26	3·5
,, 45ᵐ	65	65	91·5	77	26·5	12
4ʰ	64·5	66·5	86·5	84	22	17·5

(101.) It will be seen, on examining the above Table, that on the 15th, previous to the eclipse commencing, the temperature of the air was 4° higher, and the radi-

ating thermometer 8° higher than on the day previous.
As might be expected, the radiating thermometer was
sooner affected by the eclipse than the other, beginning
to fall almost simultaneously with the beginning of the
eclipse, whereas the temperature of the air did not fall
till nearly half an hour after the eclipse had begun.   The
extreme limit to which the air was cooled was $4\frac{1}{2}$ degrees, and this was not till 10 minutes after the greatest
obscuration; the greatest fall of the radiating thermometer was coincident with this last; and the difference
between the two thermometers at this time amounted to
only 3 degrees, having been as great as $42\frac{1}{2}$ degrees just
before the eclipse began: the diminution, therefore, in
the heating power of the sun's rays was equal to nearly
40 degrees.   Probably if the radiating thermometer
had had a blackened bulb, this diminution would have
been greater, and the instrument have fallen not merely
to the temperature of the air, but possibly below it,
from the terrestrial radiation being more powerful than
the sun's rays, so reduced in number as these last were
at the time of the greatest obscuration.

   (102.) The diminution of both light and heat caused
by this eclipse was very perceptible to the senses, but,
owing to the unclouded brightness of the day, the latter
was much more striking than the former.   There was
a sensible chill in the air, somewhat similar to that
which usually occurs about sunset when the sky is clear,
and which strangely contrasted with the intensity of
the sun's rays before the eclipse began.   There was
also an unnatural appearance in the landscape around,
the light being much subdued, and objects assuming
somewhat of a yellowish tinge, though there was no
such darkness as often occurs in thick cloudy weather

when there is no eclipse at all. It is worth noting also, that, contrary to what some observers have recorded during an eclipse of this magnitude, the birds did not appear to be particularly sensible of the change, and continued in full song the *whole time*. A pair of bantam fowls, which were feeding close by, showed no deviation from their ordinary deportment and habits.

(103.) Howard has recorded observations made by him on two different occasions of a great eclipse of the sun. The first of these was on the 19th of November, 1816, when 9 digits of the sun's disc were eclipsed, and when a fall of only *one degree* in the temperature of the air was noticed*.

At the same time he refers to an observation by Dr. John Bevis, in the Philosophical Transactions, on the Solar Eclipse of April 1, 1764, when the same depression of *one degree* of Fahrenheit's thermometer took place, though in this instance the eclipse was greater.

(104.) Howard's second instance was on the 7th of September, 1820, when, during an eclipse, the magnitude of which is not stated, the temperature of the air was lowered *six degrees*; but this seems to have been partly attributable to great sheets of cirrocumulus clouds, which passed over the sun between the period of the commencement of the eclipse and that of the greatest obscuration; so that it is doubtful how much of the effect was due to the eclipse itself. "The lowest temperature was observed about seven minutes *after* the greatest obscuration," which nearly agrees with what I have recorded above†. Howard does not appear, on

* *Climate of London*, vol. ii. p. 315.

† Daniell has also recorded some observations which he made during this same eclipse: he states having noticed a de-

either of these occasions, to have made any extended observations on the diminished force of the direct rays of the sun, but, in the second instance, he just notices that, "at the time of the greatest obscuration, the thermometer being brought from under the tree, and exposed to the sun's rays, the quicksilver rose only half a degree." This shows a nearer approach to an equality between the thermometer in the sun and that in the shade than in the case of my own observations*.

pression of temperature amounting to *five degrees,* the maximum of which was *twenty-five minutes* after the greatest obscuration. —*Quart. Journ. of Sci.* vol. x. 1821, p. 136.

* *Climate of London,* vol. iii. p. 32.

CHAPTER II.

THE WINDS.

(105.) THE winds in this climate appear so variable and irregular, especially to persons who have not paid them much attention, that at first it might be thought hopeless to endeavour to bring them under any given rules or order of succession. And certainly it requires the collected observations of many years to warrant this attempt. After all, it is scarcely possible to do more than make a seeming approximation to the truth. We can only reduce the variations of many seasons to a theoretical type, towards which there is a constant tendency, though the type is seldom, and never but very imperfectly, realized.

(106.) There can be little doubt that the general causes of all winds are to be referred to the constantly varying temperature of the atmosphere, and the unequal heating of the air over different countries in connexion with the rotation of the earth. From this arises a disturbance of the equilibrium, which the several currents that set in have a tendency to restore.

(107.) The two main currents, to which all others are subordinate, are, as is well known, the S.W., having its course from the equator to the poles, and the N.E., flowing in an opposite direction, from the poles to the equator. Into the circumstances and details of these two currents, which occasion the trade-winds*, considered in all their generality, it is not my purpose to enter. They may be found treated of in any elementary work on Meteorology. My object is merely to speak of the winds in the particular locality to which my own observations have been confined. It is probable, however, that the ever-shifting winds of the temperate regions are but modifications of the above two currents, induced in different places by the relative distribution of land and water, and other local peculiarities. "Within the tropics these currents flow one over the other, but beyond the tropics they flow beside each

* Lieut. Maury has proposed a new theory of the trade-winds, which is a more extensive generalization than the one formerly entertained. He supposes that the trades on rising at the equator, instead of returning each to its own pole, as hitherto maintained, cross over to the opposite poles. By this arrangement, "the atmosphere is united in one grand whole, and every particle in its turn courses over the whole extent of the earth's surface," instead of the atmosphere being divided into two halves, which meet at the equator but never mix. He considers this theory as best explaining the well-known fact of the greater quantity of rain which falls in the northern hemisphere than what falls in the southern; the S.W. winds which supply that rain, having derived their moisture from the immense tracts of ocean in the southern hemisphere, over which they had passed as the south-east trades, before arriving at the belt of equatorial calms, and being borne thence as an upper current into the northern hemisphere.

His reasons for adopting this view of the subject, together with the probable cause, in his opinion, to which this more extensive circulation of the atmosphere is to be referred, may be seen in the sixth chapter of his work on *The Physical Geography of the Sea*.

other in variable channels; and it is the predominance
of one of these currents over the other at a given place
which determines extreme circumstances or conditions
in respect to temperature, and their contest and alternate prevalence which determines the changeable character of the weather of our own countries*." The
struggle between these two opposite winds will be most
manifest in those latitudes in which the upper current,
from being gradually cooled in its passage towards the
N., descends to the surface of the earth. The exact
part of the globe in which this descent takes place
varies with the position of the sun; the upper current
reaching the surface in summer in somewhat higher
latitudes than in winter.

(108.) It is the assumption "that there are, properly
speaking, only two atmospheric currents on which our
relations of weather depend,—a polar and an equatorial
current,"—on which rests Dove's "Law of Rotation."
And it is perhaps this law which best serves to explain
the constant changing of the wind in temperate climates.
This observer has shown how, not merely in the more
general instance of the trade-winds, but in that of any
wind setting in to supply the vacancy caused at a given
place by an ascending current of warm air, this wind,
coming at first from the N., must, if the cause of the
movement of the air continue, take continually a more
and more easterly direction, as the current flows from
places more and more northward. This is due to the
rotation of the earth upon its axis, combined with the
circumstance of the air moving slower and slower as it
passes from higher to lower latitudes. And from the

* Dove, *On the Distribution of Heat over the Surface of the*
Globe, &c., p. 19.

same causes, a wind, which at its commencement at
any given place is due S., gradually becomes more and
more westerly. After a calm is established, by the movement of the air altogether ceasing, in consequence of
the resistances it meets with from the earth's surface,
the wind may remain for a longer or shorter interval
tolerably steady; but on a renewal of the cause which
first led to the movement, the easterly wind will fall
back to N., and the westerly to S., and the same gradual
revolution will take place as before, and in the same
direction.

(109.) Hence, under the influence of the two currents above spoken of, often alternating with each other,
the wind, in these latitudes, passes round the compass
for the most part in one given direction, viz. that of the
sun; and more frequent variations will take place between S. and W., or between N. and E., than between
W. and N., or E. and S.; since the former are due to
the main exciting cause of such variations, "while the
latter point to a change of it*."

(110.) *Revolving Storms.*—When two opposite currents, such as the N.E. and S.W. above alluded to,
come into collision, they often give rise to a rotatory
movement of the atmosphere, occasioning those revolving storms, or *cyclones*, as they have been called, which
have received so much attention from Mr. Redfield and
Colonel Reid†. Hurricanes are thought by these

* For further details in explanation of this subject, treated
of in a popular manner, the reader is referred to *Buff's Physics
of the Earth*, by Hofmann, Letters 13 and 14, on "The Winds,"
from which some assistance has been derived above.

† See *Edinb. New Phil. Journ.* vol. xviii. p. 20; and *Rep. Brit.
Assoc.* 1838, Trans. of Sects., p. 21. Colonel Reid's researches

writers to be great whirlwinds, " in which the air is
carried with extraordinary velocity round a calm centre,"
at the same time that the whole whirl moves onward in
a certain direction. The area occupied by the revolving
storm varies from fifty to several hundred miles in diameter. It is smallest at first, the circle continually
expanding, as the storm advances from the place at
which it commenced, until the fury of the latter, gradually getting more and more spent, at length subsides,
and the storm disappears. In these storms the rotatory
motion of the air in the northern hemisphere is always
contrary to the direction in which the hands of a watch
go round, or contrary " to that course of turning which
in the common changes of the wind is required by Dove's
Law of Rotation : . . . . in the southern hemisphere, the
rotation is in the same direction as that in which the
hands of a watch revolve."

(111.) Whether the storms which occur in the British Islands have the same rotatory character as the
more fearful tornadoes of the West Indies, and the
similar whirlwinds which are known in some other
parts of the globe, so as to deserve the name of cyclones,
seems disputed. Professor Forbes considers that in
certain instances it has been satisfactorily shown that
they do*. Mr. Nevins of Liverpool, however, from an
examination of the storms which visited England and
Ireland during the years 1852, 1853, 1854, has been
led to a contrary opinion. Many of the circumstances
attending these storms, he observes, were in strict
accordance with the theory of cyclones, but there were

on this subject have since been embodied in a distinct work,
entitled *The Law of Storms.*

* *Rep. Brit. Assoc.* 1840, p. 111.

others inconsistent with it. And, on the whole, he concludes " that in these high latitudes storms or masses of air do progress in a uniform order from west to east (like the hurricanes of the West Indies), and that the changes in the course of the wind during a storm indicate more or less of a curved direction; but that these changes are inconsistent with any uniform rotatory character; and that, judging from the observations of these three years, the law of cyclones does not obtain in the storms which visit the British Islands *."

(112.) *Relative Prevalence of the different Winds at Swaffham Bulbeck.* — For the purpose of exhibiting these results in a more general form, especially the relative prevalence of the different winds in each month, the winds have been collected into four classes, much after the manner adopted by Howard in his work †, the classes, however, being somewhat differently divided, and his class of variable winds being altogether omitted.

The first class (N.-E.) comprises all winds blowing from points between North and East, as well as those from due E., but not those from due N.

The second class (E.-S.) comprises in like manner all winds between E. and S., and those from due S.

The third class (S.-W.) comprises the S.W. and W. winds.

The fourth class (W.-N.) the N.W. and N. winds.

(113.) The following Table shows the number of days on which the wind blew from some point in each of

* *Rep. Brit. Assoc.* 1854, Trans. of the Sects., p. 31. Mr. Russell is equally disinclined to the opinion that the rotatory theory is capable of application to the storms in our high latitudes.—*Id.* 1853, Trans. of Sects., p. 32.

† *Climate of London*, 2nd edit. vol. i. p. 74.

these four divisions of the compass respectively, over a period of nineteen years, viz. from 1831 to 1849, both these years inclusive. The days of variable winds, as already stated, are omitted, as are also the days of such complete calm, as to afford no sensible indication that the air was in motion. There are also many omissions arising from want of observations on certain days, which make it necessary that the results obtained be considered as *mere approximations to the truth*; though it is hoped that the number of years over which the observations extend may in some measure lessen the error from this cause, and that they may have a certain value notwithstanding. It may be stated, that the months of August and September are those in which the greatest number of deficiencies occur, and which are therefore the least trustworthy. June and July are likewise deficient, but to a much less extent.

TABLE I.—*Relative Prevalence of Winds in the several months for nineteen years.*

	Jan.	Feb.	Mar.	April.	May.	June.	July.	Aug.	Sept.	Oct.	Nov.	Dec.
N.-E.	99	109	124	147	141	79	58	60	63	72	71	99
E.-S.	69	54	60	48	57	47	40	35	43	58	68	62
S.-W.	230	188	190	144	176	241	216	147	132	237	218	245
W.-N.	137	127	184	164	114	131	178	140	117	146	106	134

(114.) If we take now the sum of the above numbers under each of the four classes of winds respectively, we find them as follows :—

Sum of N.-E. 1122, making a yearly average of 59.
Ditto.. E.-S. 641, yearly average .... 33·7.
Ditto.. S.-W. 2364, ditto .... 124·4.
Ditto.. W.-N. 1678, ditto .... 88·3 *.

---

* The sum of the yearly averages amounts to 305·4, leaving

By this it appears that the relative prevalence of these four classes of winds is in the order of S.-W., W.-N., N.-E., and E.-S.;—the yearly average of the S.-W. being rather more than double its antagonistic class of N.-E., more than one-third in excess above the W.-N. class, and nearly three-and-a-half times above E.-S. The W.-N. are almost exactly one-half more prevalent than the N.-E., and two-and-a-half times more prevalent than the E.-S. The N.-E. class occupy nearly an intermediate place between the S.-W. and the W.-N., and are not quite double the E.-S., which are the fewest of all.

(115.) According to Kämtz*, the N.E. and E. winds in England, taken together, rather exceed in frequency the N.W. and N.; and the S.E. and S. are represented as not falling very far short of these last, though still the lowest in the series. It is not stated to what part of England these observations refer, and it will undoubtedly be found that the order of the winds, in respect of prevalence, is very different in different parts of the kingdom; but the above results are decidedly opposed to those obtained from my own observations in Cambridgeshire, as well as to those of Howard's, in his *Climate of London*; in both which districts the preponderance of the W.-N. winds above the N.-E. is clearly marked, and the E.-S. are very considerably less than any of the others.

59·6 days in each year unaccounted for: of these (taking Howard's estimate), about 30 may be given to the class of *variable* winds; while the remainder are either days when the air was quite calm, or on which observations were not made.

* *Meteorology* (translated by Walker), p. 48, Table showing the relative frequency of the winds in different countries.

(116.) Another remark of Kämtz would probably not
be found in disagreement with my own observations
if these last admitted of strict verification in this respect. He observes, in reference to those parts of
Europe in which his own country is situated, that the
numbers expressing the relative frequency of winds
"go on diminishing from the S.W. to the North; they
then increase again and attain a second maximum at the
N.E.; the figures then again diminish to the South."
In the above Table of my own, the winds have been
grouped into four general classes, and afford no means
of showing this result: resolving, however, these classes
into the eight points of the compass, which have been
severally paired together to form them, and to which
points the winds have been mostly referred in my
journals, the relative numbers stand as follows:—

S.W.	1613.		N.E.	834.	
W.		751.	E.		285.
N.W.	1098.		S.E.	367.	
N.		580.	S.		274.

These numbers do not at first sight appear to bear
out the above statement; but the disagreement, which
consists in the N.W. and S.E. observations being so
much higher than the W. and E. respectively, is probably due to many winds having been referred to N.W.
and S.E., which strictly blew from W.N.W. or N.N.W.
in the one case, and from E.S.E. or S.S.E. in the other;
more latitude being naturally allowed to those winds
which blew from points between the four leading divisions of the compass, than to those which blew direct
from N., S., W., and E. Were this error corrected,
and the observations distributed over a larger number
of points, the series would probably be found regularly

diminishing in the way that Kämtz describes: and these double maxima, in the S.W. and N.E. quarters, are just what might be anticipated, from the circumstance of these two winds being the two antagonistic winds in mean latitudes, to which all others are in some measure subordinate.

(117.) I now return to the above Table in order to trace the general variation of each of the four classes, under which the winds are there grouped, through the year.

I begin with the S.-W. class, as by far the predominating winds in this climate. These winds, it will be observed, attain their *maximum* in December: they are still very prevalent during January : in February, March, and April, they fall off, giving place to the N.-E. class: in May they increase, and in June they appear to reach a second maximum, nearly as high as that of December : in July they diminish again, and then continue falling off to September, when they sink to a *minimum*: in October there is a considerable increase, and they go on increasing from that time to the end of the year.

(118.) Thus it would appear, if the above results are at all trustworthy, that these winds have a double maximum and a double minimum each year; the two maxima occurring in June and December, the two minima in April and September. But it must be remembered that the observations, upon which these results depend, have been less frequent during the summer months, as before stated (113), especially during the months of August and September, and that therefore the results are probably liable to much correction from this circumstance.

(119.) The W.-N. class of winds, the second in importance in this country, are more oscillating than the last, and appear to rise and fall several times during the year. They attain their *maximum* in March, fall off a little in April, and more considerably in May; rise again in June, and attain a considerable height a second time in July,—then fall and rise alternately in an irregular manner during the autumnal and winter months, having their *minimum*, according to my observations, in November.

(120.) The N.–E. class of winds go on steadily increasing from January to April, in which month they attain their *maximum*: they are very little below the maximum in May; but in June there is a marked falling off, and in July or August, between which months there is probably little difference in respect of these winds, they sink to a *minimum*: during the three autumnal months they rise again, though very slowly; in December they increase further, and during that month and January they appear to occupy a place nearly intermediate between the two extremes.

(121.) The E.–S. winds are of much less frequent occurrence than any of the others, and are very irregular. They appear to prevail most from October to March, or during the winter half of the year, keeping nearly about the same the whole of that period. In summer they fall off, attaining their *minimum* either in July or August.

(122.) The next Table is for the purpose of showing the relative frequency of the winds, when collected into the two still more general classes of Northerly and Southerly, each embracing a semi-circumference of the entire horizon. The Northerly winds will thus be the sum of those given in the first and fourth classes of

Table I., and the Southerly in like manner the sum of those in the second and third classes.

TABLE II.—*Relative prevalence of Northerly and Southerly Winds.*

	Jan.	Feb.	Mar.	April.	May.	June.	July.	Aug.	Sept.	Oct.	Nov.	Dec.
Northerly..	236	236	308	311	255	210	236	200	180	218	177	233
Southerly..	299	242	250	192	233	288	256	182	175	295	286	307

(123.) Collected in this way, it is found that, on an average, the Southerly winds exceed the Northerly by about one-twelfth. It is also found that they prevail over the Northerly seven months in the year, the exempted months being the three spring months of March, April, and May, and the two months of August and September. They attain their *maximum* in December, while the Northerly attain theirs in April.

(124.) If, in a similar way, we collect the winds into the two general classes of Easterly and Westerly, the former embracing the first and second classes of Table I., and the latter the third and fourth classes, the results appear as in the following Table.

TABLE III.—*Relative prevalence of Easterly and Westerly Winds.*

	Jan.	Feb.	Mar.	April.	May.	June.	July.	Aug.	Sept.	Oct.	Nov.	Dec.
Easterly ..	168	163	184	195	198	126	98	95	106	130	139	161
Westerly..	367	315	374	308	290	372	394	287	249	383	324	379

(125.) Taking the sum of the above numbers in each case respectively, and reducing them to a yearly average, it is found that the Westerly winds prevail greatly over the Easterly, the ratio between these two classes exceeding two-and-a-quarter to one. In every month of the year they have the superiority. The Westerly attain their *maximum* in July, and their *minimum* in September; the Easterly attain their *maximum* in May, and their *minimum* in August. On the whole, it is during summer that the predominance of the Westerly winds is greatest, and during spring that it is least.

(126.) *Prevailing Winds of each Month, compared with Howard's results.*—It has been already stated that the four classes of winds given in Table I. are not divided exactly the same as Howard's, and therefore the results which they furnish do not admit of strict comparison with his. Howard's first class extends from *North* to East, not including the latter point; that is, it consists of his N. and N.E. observations. In like manner, his second class extends from *East* to South, not including the latter; his third class, from *South* to West, the latter not included; his fourth class, from *West* to North, the latter not included. Thus in each class, the limiting points are just the reverse of mine, in respect of the one included and the one rejected. This ought not in fact to make any great discrepancy, as strictly it is only a difference of one or at most a few points eastward and westward in the comparative range. Nevertheless, with the view of ascertaining exactly how nearly his results agree with mine, subsequently to getting out the above Tables, I threw them into the same form as Howard's, and underneath I have given Howard's own statement of the relative prevalence of the different winds in each month, against which I have placed in a parallel column the results I have myself arrived at, when adopting the same divisions as his.

Howard's average Course of Winds through the several months of the year.

"In *January*, which may be regarded as the middle of winter, we have little more than a mean proportion of N.-E. winds: yet the *Northerly*, taken together, preponderate by a *fourth* of their amount over the Southerly winds."

"In *February*, the proportions of Northerly and Southerly are reversed, the latter exceeding the former by a *third*; and this principally through the falling off of the N.-E. to one-half, and the increase of the S.-W. to their highest proportion for the year."

"In *March*, the N.-E. are in greater proportion than in any other part of the year, exceeding their own average by more than a third."

"In *April*, the N.-E. winds abate somewhat of their excess, continuing still in very high

The Course resulting from my own Observations.

January.—I find also in this month the N.-E. winds about, or a little exceeding, the mean proportion. The *Northerly*, taken together, preponderate over the Southerly by about one-eighth of their amount, or the difference is only half what Howard makes it. The S.-W. class is the one that prevails most, and is nearly at its maximum for the year.

February.—My observations for this month are very different from Howard's. With me the Northerly remain the same as last, while they are the Southerly that fall off to the amount of more than a fifth: this fall shows itself in both the E.-S. and S.-W. classes, but principally in the latter.

March.—The N.-E. increase considerably upon those of the last two months, but do not attain their maximum till next. They are not so prevalent as the W.-N., which are nearly as high as at any period of the year.

April.—As already stated, the N.-E. are at a *maximum*. The W.-N., on the contrary,

proportion. This and the preceding month exhibit about the same total preponderance of Northerly winds, as January : and in both, the E.–S. class being above its average, the general Easterly direction prevails over the Westerly."

" In *May*, the Southerly winds resume the like superiority as in February. The E.–S. class is at its *maximum*. The N.–E. class having decreased for two months, is now below its average; and the W.–N., which has decreased by an uninterrupted gradation from January, is at its *minimum* proportion : the *variable* winds are at their highest amount."

" *June*.—A preponderance of Northerly winds by more than a *third*; chiefly from the return of the W.–N. class."

" *July*.—In this month the class of W.–N. decidedly prevails over the rest : the S.–W.

have fallen off. The Northerly, taken together, show a preponderance over the Southerly, but not so striking as in the two last months. The E.–S. are at a *maximum*, as well as the N.–E.; so that the Easterly direction prevails on the whole; but the Easterly and Westerly are more nearly balanced in this month than in any other in the year.

May.—In this month (contrary to Howard) I find the Northerly winds still predominating, though their excess above the Southerly is less than in any month since January. The N.–E. class is still higher than any of the other three classes, though somewhat abated from what it was in April. The E.–S. class is also high, but past its maximum. The W.–N. is very low, and very near, if not quite at its *minimum*.

June.—A preponderance of Northerly winds, but not nearly to the extent Howard describes, and less than that of last month. The N.–E. class very much abated, but the W.–N. greatly increased.

July.—My observations on the winds of this month almost exactly accord with Howard's.

is also in high proportion; the N.-E. very low, and the E.-S. at its *minimum,* having gone off for two months."

"*August* exhibits the class N.-E. at its *minimum,* and that of E.-S. but little removed from it; while the W.-N. is at its *maximum,* having increased for three months, and the S.-W. in high proportion, having increased for two months. This month has the least proportion of variable winds."

"*September.*—We have here almost a balance between the Northerly and Southerly winds. In other respects the class E.-S. (which we must remember comprehends the former *point* and excludes the latter) takes a little from the rest, and is but little short of its highest amount."

"*October.*—The winds on the North and South sides of East are very nearly equal: but the S.-W. class predominates over the whole, and with the aid of the E.-S., exceeds the North-

The W.-N. are at a *maximum,* and more prevalent than any others. The S.-W. likewise high. The N.-E. still more abated than in June; and the E.-S. very near its minimum, though not lower than last month.

August.—I find the N.-E. and E.-S. classes both very near their minimum this month. The former, perhaps, would be quite, if my observations for this and the following month were not deficient. The W.-N. is lower than in July, but still prevails over all the others. The S.-W. class is high, and next in order, but not exceeding, if equalling, what it was in July.

September.—Like Howard, I find in this month the Northerly and Southerly winds very nearly balanced. The E.-S. class, though much higher than in either of the three preceding months, falls short of the maximum by nearly a fourth. The S.-W. remains the same as in August.

October.—My results for this month are not in accordance with Howard's. I find the Northerly winds prevailing over the Southerly: this is due partly to a considerable increase in

erly winds by a fourth of the sum of the latter."

the W.–N. class, which is nearly equal to that of S.-W.; partly to a falling off of the E.–S., which, with me, in this month sinks to its *minimum.*

"*November.* — Northerly winds now predominate by a fourth of their amount; chiefly from the increase of the class N.-E.; and the proportion of *variable* is very small."

November.—I find the Northerly and Southerly winds nearly balanced, but the latter rather prevailing. Contrary to Howard, the N.-E. class is lower than in any other month of the year except September. The E.–S. is much risen.

" *December.*—The classes in this month do not depart very far from their respective averages. We have again the Northerly and Southerly almost exactly balanced; while the Westerly are nearly double the sum of the Easterly."

December.—In three out of the four classes I find the proportions not far from the averages; but the S.-W. is high, and nearly at its maximum. The Northerly and Southerly are not so nearly balanced as last month, and the former prevail: this is from a considerable increase in the N.-E. class, and to a less extent in that of W.-N. When the Westerly, as a whole, are set against the Easterly, I find, as Howard does, that the former are nearly double the latter.

(127.) On comparing the above results, in connexion with the Tables from which they are deduced *, there is found as near an agreement as, perhaps, can be expected

---

* Howard's Table will be found at p. 77 of the first volume of his work, and is calculated from ten years' observations; my own I have not thought necessary to insert, as it differs but little from that already given in Table I.

under the circumstances. Howard's observations, it is to be remembered, relate only to London and its immediate neighbourhood, while mine have been made in Cambridgeshire. His are doubtless much more complete than mine for the period to which they refer, but mine extend over a larger number of years. The months which exhibit the closest approximation are principally the spring and summer months, with the exception of May. Those in which there are the greatest discrepancies, are the months of February, October, and November. On the whole, the following general results seem fairly established by both sets of observations equally : —

That the N.-E. class of winds is the predominant class during spring* (taking collectively the three months which constitute that season), and attains its maximum in March or April; that in summer, the W.-N. class prevails over the other three classes, reaching its maximum in July or August; and that during autumn, and the first part of winter, the S.-W. class takes its turn as the most prevalent, giving place sooner or later (according to Howard in January, according to myself in February) to Northerly winds, or winds from one or other of the two quarters E. and W. of N. respectively.

(128.) There is a further agreement with respect to the particular order of variation followed by some of these classes. Thus it appears that the S.-W. and the W.-N. classes are both at a minimum in spring; the

* Throughout this work the *Spring* is considered as including the three months of March, April, and May; *Summer* those of June, July, and August; *Autumn* those of September, October, and November; *Winter* those of December, January, and February.

former class gradually increasing from that period through the summer and autumn to attain its maximum in winter; the latter rising and sinking afterwards alternately,—first rising rather suddenly in June to attain its maximum in the course of the two succeeding months, then sinking again in autumn, then again rising at the beginning of winter, and maintaining a moderate but steady elevation through December and January. The E.-S. class is shown likewise by both sets of results to have its maximum in spring like the N.-E., and its minimum in summer.

(129.) *Double Currents.*—In estimating the average frequency of the different winds in the above Tables, notice is taken only of the current nearest the earth indicated by the vane. But in a very great number of cases, especially in unsettled weather, there are at least two currents, more or less opposed to one another, at different heights in the atmosphere. These are, generally, the two great antagonistic currents of N.E. and S.W.,—more or less affected by circumstances in respect of the exact direction they take at any given time,—which are always struggling for the mastery. Sometimes one, and sometimes the other, may prevail: but if neither predominate entirely at the place of observation, they may flow, for a longer or shorter period, " one above the other, or near to each other," giving rise to variable currents in the neighbourhood of their plane of junction, the influence of which will be continually extending itself to more remote distances. In such cases there must at first " lie between them a layer of air at rest, which passes gradually on each side into the motion of the two currents. If one wind is stronger than the other, it will by degrees draw the air that was at rest into its own movement, part of the opposite weaker stream will come to rest, and will then in like manner be dragged into the motion of the stronger current, and thus the weaker will gradually give place to the stronger*."

(130.) These double currents have not, except in some few instances, been sufficiently attended to by observers, who have generally been content with noting in their registers simply the lower current as indicated by the vane. It is very easy, however, by watching narrowly the different strata of clouds at different heights, when the sky is not uniformly overcast, or uniformly clear, to notice whether the whole body of air upwards is moving in the same direction or not. I speak here only of those currents which are below the utmost elevation to which the cloud region reaches. That there are still other, and, perhaps, many currents higher up, when we consider the great height of the entire atmosphere, can scarcely be doubted, though an observer on the surface of the earth can have no evidence of their existence. Mr. Monck Mason seems to think, and his opinion is in conformity with that of Mr. Green, the well-known aëronaut, "that in this country, whatever may be the direction of the wind below, in the higher regions, that is, generally within 10,000 feet above the surface of the earth, the direction of the wind is invariably from some point between the north and west†." If it be so, this N.W. current must be quite independent of other currents, in a lower region, which occasionally prevail, to the number of

* Buff, *Physics of the Earth*, p. 227.

† *Rep. Brit. Assoc.* 1841, p. 56. Mr. R. Russell thinks this opinion is "carried too far."—*Id.* 1853, Trans. of Sects., p. 33.

two at least, having a different direction from it, as well as from each other. It must likewise be independent of still higher currents, which sometimes set in from an opposite direction, *i. e.* more or less Southerly or S.W., as indicated by very lofty cirrus clouds, which must be often at a greater elevation than 10,000 feet.

(131.) There is one state of weather in particular, not at all unfrequent, in which there is reason to believe that an upper current always exists, different from the one next the earth, though it cannot be detected by direct observation; this upper current having sometimes a Southerly, and at other times a Northerly direction. I allude to the dull monotonous weather which occasionally prevails, in all seasons, for days together, without sun, the sky uniformly overcast, and the air nearly or quite calm. Indications of this upper current are afforded by the barometer standing at a different elevation from that due to a wind blowing in the direction of the lower one, supposing the whole body of the air overhead to be moving in this last direction. It may be inferred also from the general direction of the wind previous to, and after, the occurrence of such weather. If the lower current be Northerly or Easterly, and the barometer low, especially if the general direction of the wind have previously been S.W., and return to S.W., simultaneously with the clearing of the sky, we may reasonably infer that a S.W. current was blowing up above, all the while that the current next the earth was blowing in an opposite direction. If, on the contrary, the lower current be S.W. during the same dull weather, and the barometer high, the wind having been previously Northerly, we may conclude that the upper current is still Northerly, though the lower is changed.

(132.) The double currents just alluded to are, indeed, so closely connected with the different changes of weather in this variable climate, that these latter can never be properly understood, except the former be taken into account. With the view of drawing attention to this circumstance, and pointing out this connexion, I have selected from my meteorological Journals, and given below, some of the more observable instances in which these currents have been noticed, with the exact conditions of weather under which they have occurred. Whenever the double current has been distinctly marked, it has been my practice to enter it in the daily register by writing the two opposite winds one above the other, like a fraction in arithmetic : thus $\frac{N.W.}{S.W.}$ would imply that an upper current was flowing from N.W., while a lower one was setting in from S.W. This method may be found convenient by other observers who keep a journal of the weather, if it has not already been adopted. The state of the barometer in connexion with these double currents is particularly noticed in the instances given below. Though not much alluded to in the general conclusions that follow, this subject will be dwelt on afterwards, when speaking of the variations of the barometer as affected by changes of wind.

(Ex. 1.) 1845, March 21.—Wind $\frac{N.}{S.W.}$. Barom. very high, but falling in P.M., with light breeze from S.W. The first breaking up of a long and severe frost, N^{ly} winds having prevailed before for several days. Much milder afterwards, with the wind for the three following days S.W.

(Ex. 2.) 1845, April 2.—Wind $\frac{S.W.}{S.E.}$, with high but slightly

falling barometer; (the previous day S.E.;) the day following
S.; then returning to E., with barometer again rising.

(Ex. 3.) 1845, April 17, 18, 19.—Wind $\frac{\text{N.E.}}{\text{S.E.}}$, with high but
slightly falling barometer; the next day S.E., with increase of
temperature; then returning to E.

(Ex. 4.) 1845, Dec. 29.—Wind $\frac{\text{N.W.}}{\text{S.W.}}$ in A.M., S.W. in P.M., and
W^{ly} the next day. (Weather for some days previous stormy
and unsettled, barometer very unsteady.)   Fine and very mild
on the 30th, but on the 31st weather again changeable, wind
oscillating between S.W. and N.W.

(Ex. 5.) 1846, Jan. 3.—Wind $\frac{\text{N.W.}}{\text{S.W.}}$; (on the two preceding
days N.W.;) on the 4th S.W.   Fine and frosty on the 2nd,
and morning of the 3rd, with high barometer.   The latter falling in P.M. of the 3rd, and the next morning rain with wind.

(Ex. 6.) 1846, Jan. 30.—Wind $\frac{\text{N.W.}}{\text{S.W.}}$.   Barometer moderately
high, but falling.   Sky overcast, clouds increasing till 2 P.M.,
when rain fell for nearly an hour from S.W.: afterwards wind
passed to W., and evening very mild: (previous weather changeable, after the 30th more settled.)

(Ex. 7.) 1846, Mar. 30.—Wind $\frac{\text{S.W.}}{\text{S.E.}}$, blowing stronger and
cooler in P.M.   Barometer high all day, but falling towards sun-
set.   (Weather for several days previous cold and changeable,
with N.W. winds and low barometer: probably an upper current from S.W.: on the 29th wind N., with barometer above
30 inches.)   On the 31st fine and very mild, with wind S.W.,
to which point the lower current appeared to have worked round
from S.E. through S.

(Ex. 8.) 1846, May 22.—Wind E^{ly} last night, and very cold,
working to S. in morning, and in P.M. to S.W., with great increase of temperature, but all this while an upper current from
N.E.   Barometer high and rising the whole day.   On the 23rd
the lower current working on through W. and N.W. to N.
Barometer still rising, but temperature a little fallen.   Misty
clouds in A.M., with a little light rain.   (Previous weather very

changeable and rainy : after the 23rd fine and settled, with N^{ly} winds.)

(Ex. 9.) 1846, Dec. 1.—Wind $\frac{N.W.}{S.W.}$; severe frost the previous night, with much mist and thick rime on the trees; milder in P.M. Barometer falling from 29·878 (at 10 A.M.) to 29·628 (10 P.M.). (N^{ly} winds, with frost, some days previous as well as after, the lower current revolving alone (on the 1st), and causing the increase of temperature and deposition of moisture.) On the 2nd, wind N.W., with frost and thick mist, barometer fallen to 29·440 : probably now an *upper* current from S.W., this having changed directions with the lower.

(Ex. 10.) 1846, Dec. 23.—Lower current working in A.M. from S. to E. and N.E., and in P.M. to N., upper current all day S.W. Barometer very low, only 28·746 at 10 A.M., but rising in P.M.; sky mostly clouded, with several hours of rain. (Wind the previous day first N.W., then S.W.; the day after E^{ly}, with snow.)

(Ex. 11.) 1847, Jan. 3.—Wind $\frac{S.W.}{S.E.}$, with falling barometer and relaxation of frost, after several days of severe weather and E^{ly} winds. Barometer previously very high. Wind, on the 4th and 5th, S.W., with confirmed thaw, the lower current having worked gradually round through S. and S.S.W.

(Ex. 12.) 1847, Jan. 19.—Wind $\frac{N.E.}{W.N.W.}$: dull frost, with high but falling barometer. Jan. 20.—Wind $\frac{N.E.}{S.W.}$: sharp and very white frost, with clear sky : heavy fog-bank rising in S.W. towards sunset : barometer continuing to fall slowly throughout the day. (Steady frost for several days previous with E^{ly} winds.) On the 21st a heavy fall of snow with the wind S^{ly}, followed by a gradual thaw.

(Ex. 13.) 1847, Jan. 29, 30.—Wind $\frac{N.W.}{S.W.}$. For two days upper and lower current different, but each maintaining its own direction. Frost both days, very white on the 30th, with misty clouds. Barometer low, but gradually rising the whole time. (Wind for more than a week previous W^{ly} or S.W., with changeable and wet weather: barometer very low.) After the 30th, wind passing to N. and N.E.: barometer still rising, and more settled frost.

(Ex. 14.) 1847, Feb. 4.—Wind $\frac{\text{N.E.}}{\text{N.W.}}$. Fine and frosty, but sky lightly clouded: barometer 30·146 at 10 A.M., rising at night to 30·191. Feb. 5.—Still two currents, but upper one now N.W., while lower is W^{ly}, working on to S.W. The revolving of the lower one since yesterday through one quarter, has dragged the upper one with it through the same space. Barometer 30·147 at 10 A.M., falling to 29·944 before night. Frost in the morning: temperature rising in P.M., with misty clouds and a little rain. Wind on the 6th N.W. above and below, inclining to W^{ly} in P.M. Barometer falling to 29·520, and much milder, with light showers.

(Ex. 15.) 1847, April 24.—Wind $\frac{\text{N.E.}}{\text{S.W.}}$. Fine spring day: barometer at 10 A.M. 29·993, falling a little towards evening. (Wind the two previous days N.E., with barometer above 30 inches.) On the 25th wind W^{ly}; barometer continuing to fall, but still very fine: on the 26th S.W. with rain; barometer fallen to 29·739.

(Ex. 16.) 1847, May 11.—Wind E. in A.M., with fast rain for several hours: barometer 29·591, fallen since yesterday: sky clearing in P.M., wind at same time working through N. to N.W., and finally to S.W. Apparently an upper current from S.W. the whole day. (Wind for several days previous W^{ly} and S.W., also S.W. the day following (12th), with barometer risen, and weather more settled.) In this instance the lower current appears to have revolved alone, the upper one having its direction unaltered.

(Ex. 17.) 1847, June 7.—Wind $\frac{\text{N. to N.W.}}{\text{S.W.}}$, after a fortnight's very fine settled weather, with very high barometer, falling gradually for six days previous. (Heavy clouds occasionally, with drops of rain on the 6th.) On the 8th a decided change of weather: wind W^{ly} in A.M., with steady rain, and barometer much fallen; working to N.W. in P.M., with showers and thunder. Thence to the end of the month changeable, with much rain at intervals; wind oscillating from S.W. to N.W., sometimes $\frac{\text{N.W.}}{\text{S.W.}}$, at others $\frac{\text{S.W.}}{\text{N. or N.W.}}$.

(Ex. 18.) 1847, July 10.—Wind $\frac{\text{N.W.}}{\text{S.W.}}$, with high and rising barometer. Air close and sultry, with high dew-point and misty clouds; the latter increasing in P.M., occasioning showers, lower current at same time shifting to N.W. Temperature very high, and rising higher for several days in succession afterwards. Wind oscillating from S.W. to N.W. (Previous weather very fine, with W^{ly} and S.W^{ly} winds.) On the 14th wind $\frac{\text{S.W.}}{\text{N.W.}}$, afterwards passing to N.E., barometer continuing steady at about 30·20 inches.

(Ex. 19.) 1847, Aug. 27.—Wind $\frac{\text{N.}}{\text{S.W.}}$ in A.M., shifting entirely to S.W. in P.M., with misty clouds forming, and increase of temperature. Barometer 30·272 at 10 A.M., slightly falling, but rising again after sunset, sky at the same time clearing. (Wind N^{ly} for four previous days.) On the 28th wind N.W.; on the 29th $\frac{\text{S.W.}}{\text{N.W.}}$, with a good deal of rain; barometer still high (30·147), but falling.

(Ex. 20.) 1848, Jan. 28.—Wind $\frac{\text{S.E.}}{\text{N.E.}}$ (after several days of severe frost) in P.M. S^{ly}, with snow-showers, temperature slightly rising towards night. The following day wind due S., working to S.W. in P.M., and much milder: barometer falling. On the 30th wind S.W., and a confirmed thaw. (During the frost wind had been S.E., and barometer above 30 inches; by the evening of the 30th the latter had fallen to 29·235.)

(Ex. 21.) 1848, March 8.—Wind in A.M. $\frac{\text{N.E.}}{\text{S.W.}}$, in P.M. W^{ly}. Slight frost, with dull clouded sky, and high barometer, falling, however, towards evening. (Previous weather changeable, with wind oscillating between S.W. and N.W., and barometer below mean: on the 7th wind veering from N.W. through N. to E., barometer rising to 30·211.) On the 9th, and for some days after, wind oscillating as before, weather again unsettled, and barometer very low.

(Ex. 22.) 1848, Nov. 5.—Wind $\frac{\text{N.}}{\text{S.W.}}$, in P.M. W^{ly}, during changeable weather, with barometer below mean. On the 6th and 7th wind N.W., with clouded sky, barometer still low, but temperature risen: probably a Southerly current above.

(Ex. 23.) 1848, Nov. 15.—Clear and fine, with slight frost: wind N., in P.M. $\frac{\text{N.}}{\text{N.W.}}$: barometer at 10 A.M. 30·464, slightly falling towards night. Nov. 16.—Fine, as yesterday, with sharp frost in early A.M.: wind $\frac{\text{N.W.}}{\text{S.W.}}$: milder in the evening; barometer gradually falling all day. (Weather, for a week previous, generally fine, with frost and N. or N.E. winds: barometer very high.) For several days after the 16th, wind W^{ly} or S.W., and much milder, and barometer considerably fallen. In this instance the lower current first veered from N. to N.W., the upper remaining stationary; then the lower moved on to S.W., the upper shifting to N.W.; lastly both became W^{ly}, or oscillated from W. to S.W.

(Ex. 24.) 1848, Dec. 18.—Wind $\frac{\text{S.W.}}{\text{S.E.}}$, in P.M. due S.: fine morning, with barometer at 29·883; evening cloudy, barometer falling to 29·800. (Previous weather very mild, with S.W^{ly} winds.) On the 19th wind S.E., barometer rising before night to 30·070; weather clear; afterwards passing on to E. and N.E., with very high barometer, and several days of sharp frost.

(Ex. 25.) 1849, Mar. 29.—Wind $\frac{\text{S.W.}}{\text{E. to S.E.}}$: very low barometer, with thick misty clouds and a little rain. (Dull cold weather for a week previous, with E^{ly} winds and snow occasionally: barometer gradually falling ever since the 21st: probably an upper current from S.W. all this time.) On the 30th wind at first S.E., afterwards working to S.W. through S., with rain and thunder: temperature risen, but barometer still lower than on the 29th. S.W^{ly} winds and mild weather for several succeeding days.

(Ex. 26.) 1849, May 27.—Wind $\frac{\text{S.W.}}{\text{N.E.}}$: high and rising barometer: weather calm and fine, but much cloudiness and mist on
and off through the day. (Fine summer weather the three
previous days, with W^{ly} winds.) On the 28th wind still $\frac{\text{S.W.}}{\text{N.E.}}$,
with several hours of steady rain, and barometer still rising.
On the 29th wind N., sky clear, and barometer risen to 30·214.

(Ex. 27.) 1849, June 4.—Wind $\frac{\text{S.W.}}{\text{S.E.}}$, during fine hot summer
weather, with cirrus and cirrocumulus clouds; in P.M. S.W.,
with increased cloudiness and a few drops of rain. (On the
previous days of the month wind at first N.W., afterwards
working through N. to N.E. and E.) On the 5th wind S.S.W.,
shifting in P.M. to $\frac{\text{S.W.}}{\text{N.W.}}$, with thunder and a little rain. Barometer 30·260 on the morning of the 3rd, falling from that time
to the morning of the 5th, and then standing at 29·819.

(Ex. 28.) 1849, Aug. 6.—Fine summer's day, hot in A.M.,
wind $\frac{\text{N.W.}}{\text{S.W.}}$: towards evening both currents working N^{ly}, air
cool and misty, with rising barometer. (Wind on the 5th at
first E^{ly}, afterwards working through S.E. and S. to S.W.) On
the 7th the same two currents as on the 6th, with threatening
clouds, barometer falling. On the 8th hot and sultry, wind
S.W., barometer fallen still lower, with heavy thunder-storm
in P.M.

(Ex. 29.) 1849, Aug. 19.—Very fine and settled, with N.
wind, and high barometer, rising from 30·181 to 30·290 during
the day. Aug. 20–22.—For three days sky completely overcast, though very warm and calm, with wind S.W., and barometer standing at 30·339 on the morning of the 20th, gradually
falling to 30·145 by the evening of the 22nd. Undoubtedly an
upper current from N. or N.E. still prevailing. On the 23rd
wind again N., and sky clearing.

(Ex. 30.) 1849, Sept. 27.—Wind $\frac{\text{S.W.}}{\text{S.E.}}$, the lower current
blowing fresh, with cirrocumulus clouds and falling barometer:
much rain the night following. On the 28th wind S.W., and
barometer fallen lower. (Wind on the 26th E^{ly}.)

(133.) On a review of the above examples of two currents, it will be seen that in most instances they are attendant upon a change of weather in some one or other of its circumstances, or a tendency to change, which may or may not be carried into effect. In fact, it is this continual shifting of the wind, and constant intermixing of currents of different temperatures, when the whole body of the air at one place is not either stationary, or advancing steadily in the same direction, that causes an alteration of weather. In the examples numbered 15 and 17, we have instances of a change from fine to wet, the S.W. current first setting in beneath or next the earth, while the upper is still Northerly. In Ex. 30, previous to rain coming on, the S.W. current is found uppermost. Examples 6, 7, and 8, are those of a change from wet to fine, or more settled weather. Here too we have the S.W. current either above or below, and in one of the former cases apparently prevailing several days before the change manifests itself. In Ex. 8, where the upper current is Northerly, the lower is observed to revolve gradually, in the same direction as the sun (much more frequent than a revolution in the opposite direction), nearly round the compass, until it again fall in with the upper one. Examples 1, 5, 11, 12, 20, 23, are all instances of double currents previous to the breaking up of frost, and the coming of a thaw more or less lasting. In these instances the S.W. current appears to set in first more frequently below than above. In those in which the warmer current is uppermost, the lower one is observed gradually working from Easterly to S.W., through the S., to join it. In Examples 12 and 23, while the upper current is N.E. or N., the lower is seen to pass to S.W. in the opposite direction, or through N.W. Examples 13 and 24 are previous to the setting in of frost, after changeable or mild weather in winter. In Ex. 13 two currents are observable for two days, each maintaining its own direction (the warmer one from the S.W. being undermost, and occasioning *white* frost), before the wind passes to the N. and N.E., through N.W., and the cold becomes more settled. In Ex. 24 we find the lower current first showing itself in the direction of S.E., the move taking place through the S., and both currents afterwards passing on to E. and N.E., in the direction contrary to the sun, which, as above observed, is not very frequent.

(134.) The above are instances of double currents, attendant upon a change of weather more or less lasting. Sometimes, however, they occur just to interrupt the steadiness of the weather for a few hours, or it may be for a day, after which, things return to their former condition, and the weather goes on as before.

Thus we have an interruption of fine weather in Examples 18, 26, and 27, arising from a shift either of the upper or lower current; the interruption not continuing long, but inducing more or less of cloudiness and rain while it lasts: or an interruption of frost in winter, as in Ex. 9; here the tendency to change shows itself in a temporary increase of temperature, accompanied by much mist and deposition of moisture, if it does not lead to actual rain or snow, as is often the case. Or during wet and changeable weather, when the wind has been for some time Westerly, a brief suspension of rain takes place, the upper and lower currents separating for a time and passing off in opposite directions; as in Examples 21 and 22.

Ex. 10 is an instance of the lower current, during unsettled weather in winter, working against the sun to N.E. and N., the upper one remaining stationary in S.W., previous to much rain and snow.

(135.) It is clear, from all the above cases, that there is no fixed rule in respect of the warmer or the colder current being uppermost: each in its turn occupies either the higher or the lower position. By the *warmer* current, is meant one having its direction more or less inclining to S.W. or Southerly; by the *colder*, one more or less Northerly. Where one of the two currents is either from the N.W. or S.E., the same may be the colder or the warmer, according as the second current has its direction more or less to the S. or to the N. of this one, respectively. That current will generally in the end prevail, which is most opposed to the direction of the wind (if that direction had been for any time tolerably steady), previously to the double current setting in; or if it does not remain where it is, it will continue to shift onwards in the same direction in which it first moved, the other following after a longer or shorter interval (Ex. 14). Sometimes, however, the lower current will shift to a different point of the compass and then return, the upper all the while remaining stationary, as in Ex. 3; or it will perform an entire revolution of the compass under the same circumstances, as in Ex. 16.

(136.) Sometimes, during fine settled weather, with Northerly or Easterly winds, an upper current sets in from S.W., while the lower advances gradually to meet it through S.E. and S.; the result is an increase of temperature; and after a day or two, both currents fall back to E. (Ex. 2.)

(137.) In summer, during fine weather, two currents

may be the precursor of heat, with or without rain, dependent on the state of the barometer, and the amount of vapour in the air (Exs. 18, 19). If the air is sultry, the dew-point high, and the barometer falling, the result may be a thunderstorm, the double current sometimes prevailing for two days or more, previously to its coming up, as in Ex. 28.

(138.) During changeable weather, when the wind is often oscillating between S.W. and N.W., there will be occasionally two currents for a time, one or the other of the above winds soon prevailing again, as in Ex. 4; or the upper and lower will often reverse, and each take the direction of the other, these alternations recurring several times in succession (Exs. 9, 17, and 19).

(139.) It has been before observed that two currents probably often exist, within a moderate elevation above the earth, when the upper cannot be observed owing to the cloudiness of the sky (131). Such is conjectured to be the case in Ex. 29, where for three days the lower current is S.W., with the sky completely overcast, the wind before and after being N., and the sky in part clear. Probably the N. wind prevailed above the whole time, and the cloudiness is caused by that and the South-westerly current below intermixing; as soon as the latter returns to its former direction, the thick vapours are again dispersed, or collected into clouds of a definite form. In Ex. 25, after a week of dull cold weather, with easterly winds, and a falling barometer, during which an upper current from S.W. is supposed to have prevailed, the two currents at length become apparent, the colder one gradually passing first to S.E., and then S., and ultimately to S.W., to join the warmer one above.

(140.) *South Winds.*—The wind in this country seems

seldom to remain in the south many days together. More generally, when in that point, it is merely on its passage to some point beyond; or it quickly returns to where it was before, not continuing due S. above a few hours. Its veering to that quarter is indicative of a change of weather, and the direction may be either with or contrary to that of the sun. After a long run of fine settled weather in summer, or of cold and frost in winter, during which period the wind has been more or less steady in the E. or N.E., it may often be observed gradually shifting to S.E., and so onwards through the S. to S.W.; a change then ensues, and the weather becomes wet or changeable, accompanied, if winter, by an increase of temperature and a relaxation of frost. At other times, during the same settled weather, the wind will advance as far as S., and after remaining there a short time, instead of passing on to S.W., will run back to S.E. or E. Which of these two is likely to ensue may often be judged of beforehand by the state of the barometer: if this instrument fall much, especially if it has been falling for some time, the wind may be expected to move on to S.W.; but if it fall very little, as the wind approaches the S., the latter may be expected to return shortly to its former quarter.

(141.) When the wind passes through the S., contrary to the direction of the sun, or from S.W. to S.E., and thence onwards to E. and N.E., it likewise often leads to a change of weather, generally, but by no means always, of an opposite character to that above mentioned; this change, however, seldom continues long, and, after a few days, the wind will be found returning to S.W. the same way it came. There is often a considerable fall of rain just at the time of passage each way, and if in the summer, and the temperature high, thunder-storms not unfrequently occur at this juncture. Some-times here too, as in the instance of the wind going *with* the sun, the wind will get from S.W. to S., but after having been a few hours, or perhaps only a few minutes, in this latter point, will fall back again to S.W., without going on to S.E. This is very common in wet weather, or after rain has commenced in the W. or S.W.; and the rain increases as the wind advances towards the S., diminishing again as it falls gradually back to its former position.

(142.) If the wind *does* continue steady in the S. for any number of days together, as sometimes happens during fine weather, the barometer will be found high, probably above 30 inches, and the inference is that there is an upper current all the while flowing from the N. or N.E.

(143.) *Easterly Winds.*—The Easterly winds, which prevail in this climate in spring, for the most part blow-ing with great regularity, are known to every one. It is these winds which, from their keenness, are so trying to the constitution, and render the spring season more prejudicial to a certain class of invalids than any others. It has been already shown, in the Table of the relative frequency of the different winds in each month, that the Easterly (under which name I here include the whole of the N.-E. class, as defined in that Table) keep on increasing in frequency from the commencement of the year to April and May, in which months they attain their maximum, and then rapidly fall off. This is simply, how-ever, their average order of advance, deduced from the observations of a long term of years. It will be found that in certain particular years they occur very irregularly, and either scarcely blow at all for any continuous period, or alternate with the Westerly winds from the opposite quarter, just in a contrary order to what may be considered as the general rule. In some years the total amount of Easterly winds is full double what it is in others. They may also be chiefly confined to some one season, or distributed equally over the four seasons, or even over all the several months. It is this irregularity of the winds which conduces so much to the irregularity of the seasons in this climate; inasmuch as, the Northeasterly being for the most part dry, and the Southwesterly wet, a great prevalence of either of these classes of winds during any one year, or any one season, in particular, may quite alter its usual character, and render it colder or hotter, wetter or drier, as it may happen. At the same time, there being generally to a certain extent a balance between these two classes of winds in the course of the year, the character of a season will often be dependent on that of the preceding one, so far as the winds are concerned. It becomes interesting, therefore, to notice whether the Easterly winds occur at their right time in the spring months or not, with a view to determining the influence they may possibly have upon the succeeding summer. Accordingly I have endeavoured below to trace this connexion, so far as it exists, by noting down to what extent Easterly winds have prevailed each year of the nineteen over which my observations reach, appending such remarks respecting the weather of that year as seem to afford ground for any conclusions.

1831. — Easterly winds rather in *excess*, the greater part occurring at the regular time in the months of March, April, and May; but many likewise in July and August. *Summer* and *autumn* both moderately fine, the latter with a high mean temperature.

1832.—About a *mean* proportion of E^{ly} winds during the year; occurring principally in the first five months, but at intervals afterwards, especially in Oct. and Nov. *Summer* very fine: *autumn* fine and dry, with a temp. above mean; followed by a mild *winter*.

1833.—E^{ly} winds about the *mean*, occurring mostly in Jan., March and May; but at intervals throughout the summer and autumn months. *Both these seasons* fine and dry on the whole: following *winter* very mild.

1834.—A *mean* proportion of E^{ly} winds, most prevalent in April, but distributed over all the other months in the year, except January, which was extraordinarily mild, with the wind entirely S.W. or W^{ly}. Both *summer* and *autumn* very fine and hot; followed by a very mild *winter*. The driest year of the nineteen.

1835.—E^{ly} winds rather *under the mean*, distributed through all the months in the year, except February and September; most prevalent in March, June, July and August. *Summer* very hot and dry: *autumn* cold and wet.

1836.—E^{ly} winds *very much under the mean*, very prevalent in May, the only other months in which they occurred to signify being Feb., Aug., Sept. and Dec. Both *summer* and *autumn* changeable, with an excess of rain; the latter season cold, and winter setting in early.

1837.—E^{ly} winds this year *nearly at a minimum*, most prevalent in June and July, the rest distributed over all the remaining months, the spring and autumn having nearly an equal share. *Summer* rather changeable: *autumn* very fine: following *winter* very severe.

1838.—E^{ly} winds *rather in excess*, occurring mostly in the early part of the year, especially in Jan., Feb. and May, but prevalent also in November. *Summer* changeable: *autumn* moderately fine, with the quantity of rain below the mean.

1839.—E^{ly} winds *scarcely exceeding those of* 1837, and confined chiefly to the first half of the year; most prevalent in April and

May; none, or scarcely any, in July, Aug. and Sept. Both *summer* and *autumn* changeable and very wet.

1840.—E^{ly} winds at a *maximum*, by far the greater number occurring in the first five months, especially Feb., March and April; none in June and July, but some in autumn, and the number in Dec. equalling that of March. Both *summer* and *autumn* (excepting Aug. which was fine) rather changeable, with frequent rain. Following *winter* very severe.

1841.—E^{ly} winds at a *minimum*, occurring principally in Feb. and April; scarcely any afterwards till Oct. The wettest *summer* and *autumn* recorded in my observations.

1842.—E^{ly} winds *much in excess*, prevailing especially in April, and in less quantity throughout the summer and autumn months, with the exception of Aug. Scarcely any in Feb. and March. *Summer* very fine, with a high mean temp.: *autumn* changeable; followed by a very mild *winter*.

1843.—A *mean* proportion of E^{ly} winds, occurring chiefly in the first six months, and especially in Feb., March and June. *Summer* wet and changeable: *autumn*, with the exception of September, cold and changeable likewise. Following *winter*, the early part of it, very mild.

1844.—E^{ly} winds nearly at a *maximum*, prevailing greatly in May and December, a few in each of the other months: the two half-years about balanced in respect of their frequency. *Summer* on the whole fine, and very dry till after Midsummer: *autumn* changeable; followed by a severe *winter*.

1845.—E^{ly} winds *under the mean*, blowing chiefly in March, April and Sept.; but a few in all the other months, except Aug. and Oct. *June* fine and hot, but *July* and *August* very cool and changeable. *Autumn*, on the whole, very fine and dry; followed by a very mild *winter*.

1846.—E^{ly} winds *above the mean*, but hardly any occurring before April, and not many in that month; very prevalent in June, August and September, and a few in each of the succeeding months, especially Dec. *Summer* extremely hot and fine: *autumn* likewise fine, excepting Oct., with high mean temp.: following *winter* very severe.

1847.—E^{ly} winds *very much in excess,*—as many occurring in the second as in the first half of the year, and distributed over all the months, but most prevalent in Feb., March, April, July and Oct. *Summer,* on the whole, fine and seasonable : *autumn* rather changeable, but very mild.

1848.—E^{ly} winds *nearly at a minimum,* not more prevalent in spring than in autumn, and distributed over all the months, except Feb., June and August. *Summer* and *autumn* both very wet and changeable, followed by a very fine and mild *winter.*

1849.—E^{ly} winds *rather in excess,* and nearly as many in the second half of the year as in the first, distributed over all the months. *Summer* very fine : *autumn* at first wet and changeable, but the latter half very fine ; followed by a very severe *winter.*

(144.) On reviewing the characters of the several years above given, it will be observed that scarcely any inference can be drawn from the *total amount* of Easterly winds that may occur in any one year. These winds may be in excess, even to a maximum, or at a mean, or under the mean, and yet the summer may be either fine and settled, or wet and changeable. The same may be said of the autumn, as also of the summer and autumn taken together. No instance, however, occurs of a very fine summer, in which the amount of Easterly winds is *much under* the mean. But more importance attaches to *the mode in which these winds are distributed* over the respective seasons. It will be found that, in all cases of a fine and dry summer, either they are entirely out of place, occurring in summer and autumn instead of spring (as in 1846, the finest and hottest summer in the above series), or else, though occurring chiefly in the spring, they still keep their ground in some measure, returning at intervals during the succeeding months. In fact, it is mainly due to the wind settling itself again in the dry quarter, such as the N.-E. usually is, that the weather remains steady. On the other hand, in those years which are distinguished by wet and changeable summers, it may be observed that the Easterly winds preponderate greatly in the spring, or during the first half of the year, if they are not almost confined to that season. The only years in which there are marked exceptions to this rule are those of 1837 and 1848; but these years are both characterized by the Easterly winds being nearly at a minimum; consequently, though distributed over a wider period of time, very few fall to the lot of any month in particular. In 1841, the wettest year I ever recorded, not only were the Easterly winds actually *at* a minimum, or less in total amount than any other year in the series, but these winds were almost entirely confined to February and April, and between this latter month and October there were none at all.

(145.) Evidently, therefore, the greater or less prevalence of the Easterly winds materially aids in characterizing the *seasons*; and though we may not confidently affirm, because they have already occurred in spring, they will not return in summer or autumn, yet if they have been *much in excess at the former season*, we have reason to expect South-westerly winds (which are generally attended by more or less of wet) in the succeeding months, to restore the balance; as also, if there is a great absence of Easterly winds in spring, when we usually look for them, the chances are they will prevail later in the year, occasioning either a fine summer, or a fine autumn, or both.

(146.) The same balance is sometimes struck between one *year* and another. Thus 1837, 1839, and 1841, are all years in which the Easterly winds are either at, or nearly at, a minimum; the alternate years, 1838, 1840, and 1842, being those in which they are more or less in excess, or at a maximum. At other times, two or more years follow together, of the same character in respect of these winds, before the balance turns.

(147.) The severity of the winter depends a good deal upon the prevalence of Easterly or Northerly winds in December and January; but I can trace no connexion between the character of this season, and the amount of Easterly winds, or the periods of their occurrence, in the year preceding.

(148.) *Equinoctial Winds.*—About the times of the vernal and autumnal equinoxes strong gales of wind often set in, which, from their particular occurrence at those seasons, are termed the *equinoctial* gales. I do not find, however, that these winds return with any great regularity. In certain years the weather is perfectly calm and steady at such times; and though there may be strong winds at some other periods, both before and after the equinoxes, these seem to have no obvious connexion with the period in question. When these gales do occur, whether exactly at the time of the equinox or not, they are generally from the W. and S.W., and appear due to a sudden descent of the upper current, which is always setting in from the equator to the poles. With regard to those which take place in the spring, usually about the end of March or beginning of April, —it was supposed by the late Mr. Gough, of Kendal, that they are antecedent to that more settled state of weather during which the Easterly winds are so prevalent*. In order to ascertain how far this is the case, or whether the equinoctial gales of spring or autumn are in connexion with any other marked features of

* See *Meteorological Transactions,* 1839, vol. i. p. 67.

weather, I have thrown together in the following Table
the circumstances under which these winds have occurred, when they have occurred at all, in the several
years of the series so often before spoken of.

(*Spring.*)	(*Autumn.*)

1831.—Strong W^ly winds Mar. 12–17. Previously to their setting in, as well as afterwards, wet and unsettled weather, with E^ly and N.E^ly winds.

1831.—*No winds at or near the time of the equinox.* Both Sept. and Oct. fine and settled.

1832.—*No equinoctial winds.* A long period of fine and dry weather (from Feb. to middle of April) with E^ly winds.

1832.—First half of Oct. stormy, with W^ly winds: before and after fine and settled.

1833.—Stormy and unsettled throughout Feb. with strong W^ly winds: followed by keen E^ly and N.E^ly winds in March.

1833.—*No equinoctial winds*: Sept. and a great part of Oct. fine and settled.

1834.—Strong W^ly winds at intervals during March: April fine and dry, with E^ly winds.

1834.—*No equinoctial winds.* Great part of Sept. and first half of Oct. very fine.

1835.—Heavy gales from W. end of Feb. and first week in Mar. Preceded by changeable and unsettled weather, with W^ly winds: the same weather and the same winds prevailing also for a short time afterwards, the latter then giving place to N.E^ly.

1835.—Strong W^ly and S.W^ly winds at intervals during September. Previously a long run of fine hot weather: followed by much wet.

1836.—Stormy and unsettled at intervals during March and April, but no very heavy gales. Followed by fine weather in May, with E^ly and N.E^ly winds.

1836.—Strong wind, with storms, second week in Oct. Much changeable weather previously: last fortnight in Oct. more settled, but cold.

1837.—*No equinoctial winds.*

1837.—A change from wet

Very cold March and April; severe weather breaking up on the 21st of the latter month, without wind.

1838.—Strong winds at intervals during each of the first four months, but *none in particular at the time of the equinox*. E^{ly} winds on and off the whole period.

1839.—W^{ly} winds blowing at intervals during March, but no very heavy gales. E^{ly} winds also on and off, with a long continuance of cold changeable weather.

1840.—Blowing fresh from N.E. at the time of the equinox, but no heavy gales, and no marked change of weather ensuing. The same winds prevailing before and after.

1841. — Equinoctial winds blowing from S.W. for nine days in succession towards the end of March. (Fine and mild weather had set in a few days previously, after a severe February with great prevalence of N.E^{ly} winds.) Followed by changeable weather with shifting winds.

1842. — Equinoctial winds from S.W. and N.W. the latter half of March. Attended by a change from mild to cold and stormy weather with E^{ly} winds.

to fine and more settled weather just at the time of the equinox, but unaccompanied by any strong winds.

1838.—Stormy, with gales of wind, the last week in Oct· Previous part of that month wet: Sept. fine and settled.

1839.—Strong gales from N· and N.E. the end of Oct., but none earlier. Preceded by a long run of wet and changeable weather with W^{ly} winds: the same weather returning after the gales.

1840.—*No equinoctial winds.* Greater part of September and October changeable.

1841.—Strong S.W^{ly} winds the end of September. Preceded as well as followed by wet and changeable weather.

1842.—Strong winds from E. and N.E. the end of Sept. Previously mild and changeable: afterwards fine and dry with N.E. winds.

G

1843.—Strong winds from E. and S.E. the last week in Mar. Preceded by ten days of very fine and mild weather with W^{ly} winds: followed by moderately fine, with var. winds.

N.B. Very stormy before, on Feb. 3 and 4, with strong winds from N. and N.W., previously to very severe weather setting in, and the wind fixing itself in N.E.

1844. — Equinoctial gales from W. and N.W. blowing for ten successive days the middle of March. Preceded by a long run of stormy and ungenial weather: followed (but not immediately) by change to mild spring weather.

1845. — Equinoctial gales from W. the last week in March. Preceded by nearly two months of very severe weather; followed by milder and more genial; but a prevalence of E^{ly} winds both before and after the gales.

1846.—W^{ly} winds blowing rather fresh at intervals during March and April, accompanied by changeable weather: no heavy gales, and very few E^{ly} winds previous to the last week in May.

1847. — Equinoctial gales from S. and S.W. blowing for several days together the middle

1843.—Strong gale of wind from W. and S.W. Oct. 7 and 8. Preceded by several weeks of very fine and mild weather with E^{ly} winds: followed by cold and wet.

1844.—Strong W^{ly} winds Oct. 2 and 3. Fine autumnal weather previously, with prevalence of E^{ly} winds: wet and unsettled afterwards, to the end of October.

1845. — Equinoctial winds blowing strong from W. and S.W. the 3rd week in Sept. Preceded by very fine weather with E^{ly} winds; followed by changeable.

1846.—Strong gale from W. and S.W. Sept. 24. Very fine weather previously with N.E. wind: wet and stormy October.

1847.—W^{ly} and N.W^{ly} winds blowing fresh at intervals throughout Sept. and Oct.

of March. Preceded by severe weather with N. and N.E. winds: afterwards a change to mild, and like spring.

N.B. Heavy W^{ly} gales before the 3rd week in Feb., during a short thaw in the middle of frost.

1848.—W^{ly} winds blowing fresh at intervals during Feb. and March, especially the last week in Feb. Both these months very mild and changeable, with a great deal of rain.

1849.—W^{ly} winds blowing rather fresh on one or two occasions in March and April; but never very strong: greater part of each month very fine and mild.

Both these months changeable.

1848.—A change from fine settled weather with N^{ly} winds, the last week in Sept., to wet, with W^{ly} winds; but the latter not blowing hard.

1849.—Greater part of September and first half of October changeable, with variable winds, blowing fresh occasionally, but never very strong.

(149.) The first thing to be noted in the above Table is that, in scarcely more than half the number of years given, have any particularly strong winds occurred at all about the time of the vernal or autumnal equinox. Also, although these winds, when they do blow, are generally more or less Westerly, it will be observed that in one instance at the time of the vernal equinox they were from the S. as well (1847),—at another from the E. and S.E. (1843); likewise, at the time of the autumnal equinox, there is an instance of gales from the E. and N.E. (1842).

(150.) It will be observed further, that the equinoctial gales by no means lead in all cases to the introduction of the Easterly winds in spring. There would seem to be some connexion between them, in those years in which the Easterly winds are most prevalent (as is generally the case) at that particular season; but the latter quite as often *precede* as *follow* the gales in question; and when they do follow, they often do not follow immediately, as in the year 1835. There are also some years, as 1831 and 1845, in which the Easterly winds are prevalent both before and after the occurrence of the gales. The nearest approach that can be made to a general rule on this head seems to be as follows: if the Easterly winds prevail from an early period of the year, and do not continue for any great length of time, then, whenever they give way, the gales occur, previous to the introduction of more genial breezes: if the Easterly winds do not set in for any continuance till about the usual period, in March or April, then the stormy and unsettled weather will generally be found prevailing in the previous month, *i. e.* February or March; and may either be coincident with the time of the vernal equinox, or may take place sooner than the equinox: if the Easterly winds commence very early in the year, and continue steady till the spring is considerably advanced, there may be no equinoctial gales at all, nor any particularly strong winds before or after the equinox, as in the year 1832:—there is yet another case, viz. when the Easterly winds, without continuing to blow steadily for any great length of time together, nevertheless keep on and off, alternating with winds from the Western hemisphere, in an irregular manner; in such seasons there is generally a great prevalence of stormy weather, and strong gales recur frequently, some of which may take place at the equinox, or otherwise, as it happens. Similar changeable weather will be experienced in those years, in which there are scarce any Easterly winds till very late in the spring, as in 1846.

(151.) Thus the equinoctial gales, when they occur at all, seem rather to be attendant upon, or at least the forerunners of, a change of weather, than to serve as an introduction of the Easterly winds; though a change of the wind, more or less permanent, generally occurs at the same time, if previously it had been long fixed in any particular quarter. The equinox being about the time when a decided change of weather often takes place, gales likewise are frequent at that period. After a severe winter, or a long run of dry frost, especially, such gales generally indicate the breaking-up of the hard weather, and lead more or less directly to the commencement of spring. This was the case in 1845, and again in 1847 : in the former of these years, the gales occurred from the W., blowing for several days toge-ther in the last week of March, and raising the mean temperature of that week nearly ten degrees above that of the preceding one: in 1847, the gales were from the S. and S.W. about the middle of March, the mean tem-perature of the week in which they occurred being an advance upon that of the week previous by 13 degrees. At the same time the change is not necessarily from cold weather to milder : it is quite as often the reverse, if, from the late setting-in of Northerly and Easterly winds, the previous winter has been unusually mild. Exceptional years too occur, such as 1831 and 1843, when strong gales take place at the time of the equinox, without bringing about any lasting change of weather at all; in the latter of these two years, however, strong winds, more exactly representing the equinoctial gales in character, had taken place previously in the begin-ning of February, on the occasion of the wind fixing itself in the N.E., accompanied by frost.

(152.) In like manner, gales at the autumnal equinox often attend the breaking-up of a fine summer, bringing in wet and changeable weather, as in 1835, and several other years in the above series; or they may terminate a wet season, and usher in fine and dry weather in its place, with Easterly winds. If changeable weather prevails through September and October, accompanied by variable winds, no particular gales occur that can with any propriety be called equinoctial, though it may be more or less stormy at intervals during the whole period, as in 1847, and again in 1849 :—or a run of fine settled autumnal weather may continue a long while without any strong winds, as in 1831 and 1833. Or here also, as at the time of the vernal equinox, there may be no wind to signify, and yet a decided change of weather may take place just at the time of the equinox, as in 1837. There are other similar irregularities which can be reduced to no certain rule.

(153.) *Hot and cold winds.*—Few persons can have failed to notice a great difference in their sensations of heat and cold, according to the quarter from which the wind blows. Though the weather may remain the same in other respects, and the sun shine equally bright,— yet, if the wind shift but a few points one way or the other, its effect in raising or lowering the temperature of the air is often almost immediate. The cold biting Easterly winds of spring, which cause such discomfort to invalids, and which are so generally prejudicial to health, have been already spoken of (143). The soft Westerly winds, which prevail in the decline of the year, before frost and winter have set in, and the hot Southerly winds that occasionally blow in summer, are equally familiar to us. But it requires more exact observation to perceive that the *relative* temperatures of the different winds are not always the same, and that they vary with the season. The winds which are the hottest and coldest in summer are not the same with those which are the hottest and coldest in winter. Howard has observed (and it is quite in agreement with my own observations) that "warm weather *in winter* almost uniformly comes from the S.W., S., and W.; but *in spring and summer* from the S. and E.*" In like manner the coldest winds in winter and spring are from the N.E., but in summer generally from the N.W.

(154.) There is still, however, a *mean* temperature belonging to each wind, taking one season with another, which probably varies but little at the same place from year to year, and which it is important to ascertain. It is interesting, also, to note the order in which the winds arrange themselves, when considered in this point of view.

(155.) I have no results of my own to offer on this subject, but it has received attention from some foreign meteorologists; and Kämtz has given an abstract of observations carried on by M. Otto Eisenlohr in particular, at Carlsruhe, for 34 years, and compared with others made at Paris, London, Hamburgh, and Moscow†. I shall content myself with copying the results obtained in respect of London alone, being the only ones of much interest to our countrymen generally, and, in so doing, arrange the winds, not, as in the original table,

---

* *Climate of London*, vol. i. p. 27.

† *Meteorology* (by Walker), p. 159. The temperatures are given by Kämtz in degrees of Centigrade, but the same table will be found, in degrees of Fahrenheit, in *Thomson's Meteorology*, p. 428.

according to compass, but in the order of their respective temperatures.

Mean Temperature of the different Winds.

S. .... 52·43

S.W. .... 51·55

S.E. .... 51·04

W. .... 50·43

E. .... 49·33

N.W. .... 47·68

N.E. .... 46·54

N. .... 45·77

(156.) It thus appears that, taking all the seasons together, the S. wind is the hottest, and the N. the coldest, that blows in this country, which is what, perhaps, might be expected, seeing that the maximum degrees of heat and cold are sometimes on the one and sometimes on the other side of those two points respectively, according to season, as above pointed out.

(157.) If it excite surprise that the Easterly winds are not colder, it must be remembered that a great part of the so-called Easterly winds in spring are really more or less North-easterly, and that, though, nevertheless, the true Easterly winds are very cold during the first five months in the year, they are very hot in summer, at which season they often continue to blow much after their usual time of ceasing.

CHAPTER III.

BAROMETER, AND ATMOSPHERIC PRESSURE.

The Barometer.—Next to the thermometer the barometer is undoubtedly the most important instrument in
meteorological researches, and after inquiring into the
conditions of the temperature of any locality, it is desirable to attend to the circumstances connected with
the pressure of the superincumbent atmosphere.

(158.) The barometer, perhaps, is more generally
observed than the thermometer, on account of its indications as a weather-glass.   There are few respectable
houses which do not possess one of these instruments,—
though too often of a very inferior make,—the same
being consulted by the family every morning, to ascertain what is likely to be the weather during the day.
And, no doubt, the barometer will reveal something on
this head.   Neither can there be any question as to the
importance, in certain cases especially, of being able to
form some judgment in respect of the coming weather.
To say nothing of other classes of the community, who
are all more or less concerned in this matter, the operations of the farmer and gardener depend so much upon
season and weather, that their interests would be greatly
promoted by a knowledge of the laws by which atmospheric changes are regulated.   Yet few farmers

G 5

or gardeners possess this knowledge. They are in
general content with the traditional maxims handed
down to them by their fathers, or certain vulgar no-
tions, which possibly may have some truth in them,
but which, from their uncertainty, hardly admit of any
practical bearing. Thus I have known farmers wait for
the change of the moon before beginning to cut their
hay; or they have been guided by some other indication, which they were in the habit of interpreting as a
favourable sign, while they entirely overlooked those indications, which pointed directly the opposite way. Even
persons who look to more trustworthy sources of information about the weather, and who for this purpose
consult the barometer, seldom draw any correct inferences from this instrument, from not sufficiently understanding the language which it speaks. The mere
height of the barometer is scarcely more to be trusted
than any other indication of the weather we may be
disposed to look to singly. There may be occasions
when there is no mistaking its signs, but, in general,
unless combined with other observations, on the wind
and temperature especially, it will quite as often mislead, as guide us aright. Instances sometimes occur of
a very high barometer, yet accompanied by rain; as
also of a very low one, the weather notwithstanding
being fine. It is this circumstance which often makes
people out of humour with their barometer, and induces
them to cry it down as an instrument not to be trusted;
whereas there is no instrument more valuable for common use, if properly understood. The fault is with those
who do not know how to look at it.

(159.) Too much reliance must not be placed upon
the words *Fair, Change,* and *Rain,* which are engraved

on the plates of most barometers. It may be questioned
whether they even indicate those heights of the mercurial column, at which the states of weather severally
signified by such terms ordinarily prevail. Certainly,
however, it is often far otherwise. Without any material variation of the column, there may be steady rain
if the wind is in one direction, while, other conditions
remaining nearly the same, it may be quite fair, if the
wind is in a contrary direction. Neither is it sufficient,
in all cases, to notice the actual elevation or depression
of the barometer, for the reasons above given; independently of the circumstance that the same height of
the mercury would give a different indication in respect
of weather, according to the level of the situation in
which the observer is placed. The lower that level is,
the higher the barometer must stand, *cæteris paribus*, to
insure fine weather.

(160.) The great point to be attended to, is the rising
or falling of the barometer above or below its mean
yearly height at the locality in question. This last may
be considered, in a general way, as the turning-point.
It should be further noticed, whether it is in an oscillating state, rising and falling by turns in very short
periods of time, or whether its movements continue
steadily in the same direction, advancing or receding
by very gradual steps. An oscillating barometer always
indicates unsettled weather, when it may be quite fair,
perhaps, one hour, and hard rain the next, the sky becoming overcast very suddenly. But if from being very
low, the barometer gradually rise until it get considerably above the mean, we may then with tolerable security reckon upon fine weather shortly, though the rain
continue more or less for some time after the rise has

commenced; as, on the contrary, if from being above the mean, a gradual fall of the barometer take place to a point considerably below it, a change to wet may be almost certainly looked for, and the more certainly the more gradual the fall is, the weather, notwithstanding, continuing fine while the downward movement is going on. If to observations of this nature we add the results deduced from a knowledge of the mean height of the mercurial column corresponding to each of the eight chief winds, we have then very trustworthy *data* upon which to ground our judgment of the weather, so far as it is indicated by the barometer.

(161.) The barometer is especially valuable, as revealing to us the changes which are going on in the upper strata of the atmosphere, and which, but for the help of this instrument, we could know nothing about. Measuring the whole weight of the superincumbent atmosphere, it tells us of alterations in this weight caused from time to time by changes of temperature and wind. If the temperature on the surface of the globe were everywhere the same, the air would be everywhere at rest: its pressure, too, would be the same everywhere at equal heights, and the barometer would be invariable. But this equilibrium is constantly being disturbed by the varying influence of the sun in different regions, giving rise to currents and local displacements of the air, and consequently to fluctuations in the barometer. If the temperature in a particular locality rises, the air above that place becomes rarefied, and ascending to a greater elevation than it stood at before, flows over into the adjoining regions, by which its whole weight is diminished, and the barometer falls. If the temperature is diminished, a contrary effect is produced: the column of air having its density increased sinks to a lower level, while other air flows in from the neighbouring parts to fill up the void, increasing the pressure, and causing the barometer to rise. Thus, then, some at least of the fluctuations of the barometer would seem to be occasioned by a variation in the density of the lower regions of the air. This, however, will not explain them all; nor, especially, will it serve to account for those changes of weather, by which the barometric movements are generally accompanied. It must be further remembered, therefore, that there is an atmosphere of aqueous vapour as well as one of air above our heads, the former fluctuating in amount equally as the latter, and exercising likewise a certain influence upon the barometer. This vapour is derived from the evaporation of water from the surface of the earth, the quantity taken up depending mainly upon the temperature; and it is the precipitation of it, when exceeding in amount what can be upheld in the atmosphere at the existing temperature, which causes rain. The precipitation may either take place where the evaporation took place, in consequence of the lowering of the temperature of the air from some cause at that particular spot; or a large body of air saturated with vapour may be carried by winds into a far-distant region, and the vapour there be precipitated, in consequence of its mixing with the air of colder latitudes. In the former instance the rain probably would not be considerable, and the barometer might not be affected to any sensible amount, or might even rise while the rain is falling. If in the latter case the barometer fall, as it undoubtedly would, it is not due to the rain, as is sometimes supposed, but to the rarefied state of the air by which the rain is accompanied. The accumulation of vapour in itself would tend by its pressure to *raise* the barometric column, but such accumulation can only take place when the air has been much rarefied by heat, which last influence, being the greater of the two, makes the barometer fall.

(162.) The excess, therefore, of aqueous vapour in the air, and the fall of the barometer, instead of being one the cause of the other, are, to a certain extent, due to the same cause. Either a heated current sets in from the S.W., which is always comparatively a moist wind, from passing over the Atlantic Ocean, or, if local causes tend to rarefy the air at a particular place, the vapour of the surrounding regions flows in more freely, in consequence of the more ready admission given to its particles*. In either of these cases the barometer would fall; and the chances of rain following, if such state of things continued long, would, especially in the former case, be considerable. But in the instance first mentioned, it is easy to see that the connexion between the rain and the fall of the barometer is not a necessary one, except in reference to countries situated like our own. The circumstance of the S.W. wind bringing up vapour is due entirely to the geographical position of Great Britain. If, instead of a great ocean lying to the westward of this country, there were a continent of equal

---

* It was formerly thought that the aqueous vapour in the atmosphere was united *chemically* to the air, or dissolved by it; but Dalton has shown that the movements of the particles of the former are quite independent of the movements of the particles of the latter, and that, in fact, the union of the vapour with the several gases of the atmosphere is a mere *mechanical* one.

size, yet more if that continent consisted of an extensive sandy desert, the S.W. wind, being a hot wind, would still make the barometer fall, while, being devoid of the greater part of the vapour with which it is now charged, the air would remain clear, and there would be no rain. Nor, in the case of local rarefactions of the air, does rain necessarily follow, though the barometer fall; for the vapour in the surrounding regions may not be sufficiently in excess, or flow in sufficient quantity, to lead to a precipitation.

(163.) In like manner we may explain why the rise of the barometer ordinarily indicates the return of fine weather. Its rising is due to the setting-in of northerly currents, which, being cold, increase the density of the air, and which, being at the same time dry, take up the superabundant moisture. Yet whether the rain, which may have been falling previously to the wind shifting to the N. or N.E., cease as soon as these winds begin to blow, must depend entirely upon the amount of the vapour previously existing in the air, in connexion with the temperature of the northerly current. If the amount of aqueous vapour be very great, and the change of wind have caused a considerable reduction of temperature, the rain may continue for some time after the barometer begins to rise, or even fall heavier than before: whereas, if the amount of vapour be less, and the difference between the temperature of the southerly or westerly wind which previously prevailed and that of the northerly current be less also, the rain may cease almost immediately. In either case, however, if the northerly winds continue blowing, the fine weather will return after a time, and the barometer will continue rising till it is finally established.

(164.) From these general remarks on the falling and rising of the barometer, in connexion with changes of the weather from dry to wet and wet to dry, I proceed to give the results of the observations made with my own instrument at Swaffham Bulbeck, for the same period of nineteen years, previously adverted to under the head of Thermometer and Temperature.

(165.) The barometer employed for this purpose was one of Newman's, having an iron cistern, and a float in front to adjust the level of the mercury in the cistern to the bottom of the scale, which last is of brass throughout its length. This adjustment, which obviates the necessity of any correction for the capacity of the cistern, was in all cases accurately attended to, before making an observation. The barometer is furnished with a thermometer, the bulb of which dips into the mercury in the cistern; and the temperature was always noted down simultaneously with the height of the mercurial column. The diameter of the tube of the barometer is ·35 inch. The vernier is so divided as to read off to five-hundredths of an inch, and by the help of light admitted from behind, its index can be brought very accurately down to the convex surface of the mercury in the tube. The results obtained by this barometer have been all corrected for capillarity, and reduced to the temperature of 32°. The corrections were made according to the Tables published in the Report of the Royal Society*. The instrument was fixed in an apartment on the ground floor, with a N.W. aspect, and without any fireplace in it, so that the temperature of

* *Report of the Committee of Physics, including Meteorology; approved by the President and Council.* 8vo. London, 1840. See pages 81 and 82.

the room was not subject to any sudden changes. Care was taken at the time of fixing to insure its perpendicularity. The position of it remained unaltered during the whole period for which the observations were continued. These observations were made, as in the case of the thermometrical ones, every day at the hours of 10 A.M. and 10 P.M., with only occasional interruptions.

(166.) The first results of importance to be recorded are those given in the following Table :—

TABLE I.—*Showing the mean height of the Barometer at Swaffham Bulbeck, together with the greatest and lowest observed heights, and the greatest range, during each of the above nineteen years.*

Year.	Mean Height.	Greatest Observed Height.	Least Observed Height.	Greatest Range.
1831.	29·853	30·570	28·852	1·718
1832.	29·979	30·535	29·138	1·397
1833.	29·900	30·754	28·780	1·974
1834.	29·990	30·682	29·017	1·665
1835.	29·939	30·880	28·805	2·075
1836.	29·839	30·726	28·724	2·002
1837.	29·952	30·718	28·900	1·818
1838.	29·846	30·554	28·592	1·962
1839.	29·867	30·568	28·855	1·713
1840.	29·911	30·722	28·510	2·212
1841.	29·769	30·530	28·818	1·712
1842.	29·950	30·587	28·770	1·817
1843.	29·885	30·550	28·143	2·407
1844.	29·826	30·414	28·665	1·749
1845.	29·857	30·551	28·727	1·824
1846.	29·841	30·672	28·739	1·933
1847.	29·906	30·551	28·388	2·163
1848.	29·812	30·513	28·625	1·888
1849.	29·895	30·866	28·987	1·879
Mean...	29·885	30·628	28·739	1·890

(167.) The mean height of the barometer at Swaffham Bulbeck for the nineteen years in question is seen
by the above Table to be 29·885. This mean, however,
as already stated, is deduced from observations made at
the homonymous hours of 10 A.M. and 10 P.M., which
are near the times at which the barometer, on an average, stands highest during the twenty-four hours. It
is necessary, therefore, to make a slight correction, in
order to approximate it more nearly to the true mean,
which, according to Kämtz, takes place about mid-day,
generally between mid-day and one o'clock. It is well
known that the barometer is subject to horary oscillations arising from the influence of the sun, and quite
independent of those due to other causes. These horary oscillations vary with the latitude and the season*.
In our own country they have been well demonstrated
by some observers, especially by Mr. Snow Harris, in
a Report to the British Association in 1839†, giving
the results of hourly observations of the barometer at
Plymouth for a period of three years. From these observations it appears that the barometer is subject to
two maxima and two minima during the twenty-four
hours, the former occurring at a little before 10 A.M.,
and at 10 P.M., the latter at a little after 4 A.M., and a
little after 3 P.M. The amount of the oscillation during

* Under the tropics, where the variations of the barometer
from other causes are extremely slight, Humboldt observes that
the regularity of the *horary* variations "is such, that, in the
daytime especially, we may infer the hour from the height of
the column of mercury, without being in error on an average
more than fifteen, or seventeen minutes."—*Cosmos* (Sabine's
Transl.), vol. i. p. 308.

† *Rep. of Brit. Assoc.* 1839, p. 149.

each of the four periods is, approximately, as follows :—

Rise from 4 A.M. to 10 A.M........ ·014 inch.
Fall   „   10 A.M. to 3 P.M........ ·017 „
Rise   „   3 P.M. to 10 P.M........ ·021 „
Fall   „   10 P.M. to 4 A.M........ ·018 „

To reduce, therefore, the mean height of the barometer at Swaffham Bulbeck as given above to the true mean, it would probably require about one-hundredth of an inch to be subtracted from it. It would then stand at 29·875 inches; and it is remarkable how little this differs from the mean height of the barometer for noon at Greenwich, which, according to Mr. Belville, is 29·872 inches, on an average of thirty years*. The mean height of the barometer at the Cambridge Observatory, from 10 years' observations, commencing with 1837, and ending with 1849 (the years 1844, 1845, and 1846 being omitted), is 29·906 inches.

(168.) The highest mean during the above series of years is 29·990 inches in 1834; the lowest is 29·769 inches in 1841. The difference between the two is ·221 inch, showing the range of the mean during the same period. The above two years in which the highest and the lowest mean occurred were also respectively the driest and the wettest years in the whole series, thus showing the connexion which exists between the height of the barometer and the quantity of rain, when taken on a long average.

(169.) The greatest observed height was on January 2nd, 1835, when the barometer rose to 30·880 inches, corrected†; but it was nearly as high in February 1849,

* *Manual of the Barometer*, p. 18.
† The barometer of the late Rev. J. Hailstone stood this same

rising then to 30·866 inches. The mean annual greatest pressure is 30·628 inches; and there is only one year in the series (1844) when it was below thirty inches and a half; but it seldom gets to thirty inches and seven-tenths, and very rarely indeed to thirty inches and eight-tenths. The least observed height was 28·143 inches, which occurred in January 1843. It was also as low as 28·388 inches in December 1847, and these are the only two instances in the whole series, in which the barometer fell lower than twenty-eight inches and a half, the mean annual lowest depression being 28·739 inches. The difference between the greatest and least observed heights, or the entire range of the mercurial column during the whole period, is 2·737 inches, the mean annual range being 1·890 inch.

(170.) The above extremes are a little within those that have been observed at Greenwich. In 1825, Belville records the barometer to have stood there at 30·89 inches, being one-hundredth higher than it was at Swaffham Bulbeck in 1835; and in 1821 to have fallen to 27·99 inches: these extremes, however, are those of a period reaching over 38 years. The greatest height of the barometer I can find on record is that mentioned by Thomson in his "Introduction to Meteorology*," who states that, at 9 A.M., on the 9th of January 1820, it stood at Leith, 60 feet above the sea, at 30·999 inches; consequently, at the sea-level, it must have been above thirty-one inches; but it is not said whether this was the observed or the corrected height. Sir George Shuckburgh is recorded to have observed

day at Trumpington, near Cambridge, at 30·913 inches, but I believe this observation to have been uncorrected.

* *Introd. to Meteorology*, p. 29.

the barometer in London, in 1778, at 30·935 inches*. With regard to the greatest depression ever noticed, perhaps that recorded also by Thomson is the most remarkable : he states that on the morning of the 7th of January 1839, between 5 and 6 o'clock, the barometer was depressed at Edinburgh to 27·6 inches†. There was also a very great depression of the barometer, and one that was very generally noticed throughout the kingdom, on the 25th of December 1821. A Troughton's mountain-barometer at the Royal Observatory, Greenwich, is said to have sunk on that day as low as 27·89 inches‡. Howard records a still lower observation of 27·83 inches, about 5 A.M. on the same day, at Tottenham, with a portable barometer of Sir H. Englefield's construction§. At Trumpington, near Cambridge, this great depression was noticed by the late Rev. J. Hailstone, who made it the subject of a communication to the Cambridge Philosophical Society ; but the barometer did not fall there below 28 inches‖. Assuming that the above instances of extreme elevation and depression of the barometer mentioned by Thomson are correct, they widen the whole range within which the oscillations of the mercurial column in this country take place to considerably more than 3 inches¶.

(171.) One thing is observable, with respect to the *extremes* of the barometer as compared with the *mean* height, very different from what takes place in the ther-

* *Belv. Man. of Bar.* p. 19.
† *Introd. to Met.* p. 31.   ‡ *Belv. Man. of Bar.* p. 20.
§ *Climate of London*, vol. iii. p. 69.
‖ *Cambridge Phil. Trans.* vol. i. p. 453.
¶ Since writing the above, I find a record of the barometer having fallen at Kingussie in Inverness-shire to 27·20 inches. This most extraordinary depression took place during very

mometer; and that is, that the greatest elevations of
the mercurial column rise to scarcely more than half as
much *above* the mean, as the greatest depressions sink
below it. The whole range of the column in my series
of years being 2·737 inches, it will be found that only
·995 inch is the difference between the mean and the
greatest observed height, while the difference between
the mean and the lowest observed height is 1·742 inches.
Taking the wide extremes already alluded to as having
occurred at London and Greenwich, and comparing
them with the mean height of the barometer at the
latter place, these two portions of the entire range
are 1·063 inches and 1·982 inches respectively, the result
being nearly the same here also. In the case of the
thermometer, on the contrary, the mean of the highest
and lowest temperatures occurring at any place over a
considerable number of years makes a very close ap·
proximation to the true mean, if it be not found almost
coincident with it.

(172.) Very great elevations of the barometer, as well
as very great depressions, never occur but in the winter
half-year. Nearly half the maxima in the preceding
Table took place in the month of January; the remainder in some one or other of the six winter months. Of
the minima a larger number occurred in December than
in any other month, but, on the whole, these are more
equally spread over the different winter months than
the maxima.

stormy weather, on the 20th of November 1838, at 3 P.M. The
elevation, however, of Kingussie above the sea is, I believe, very
considerable.—(See *Rep. Brit. Assoc.* 1839, p. 28.)

The barometer at Swaffham Bulbeck that same day was
depressed to 28·592 (corrected); this occurred at 7½ʰ A.M.

The circumstances under which the two extreme heights of the mercurial column take place are, as might be expected, very different.

(173.) The greatest elevations occur during fine settled weather with northerly or easterly winds; but they are not always attended by frost, or more frost than is due simply to terrestrial radiation at night under a clear sky in winter. The rise of the mercury at these times is generally very gradual. On the occasion of the great elevation in January 1835, when the barometer rose to 30·880 inches, the instrument, except on one or two days, had stood at considerably above 30 inches for a month previous, and did not fall materially for a week afterwards. In that, likewise, which occurred on February 11th, 1849, the barometer had been gradually rising for several days previous, and had not been below 30 inches from the end of January. After it had attained its maximum of 30·866, it fell as gradually as it had risen, and did not sink below 30 inches again till eight days afterwards, having continued at or above that height very nearly three weeks.

(174.) A very high barometer *may* occur with a west or south-west wind, though ordinarily these winds have the effect of depressing the mercurial column. But whenever this is the case, there is probably an upper current in an opposite direction from some point in the northern hemisphere. Thus for several days during the last week in December 1843, the barometer ranged from 30·400 inches to 30·565 inches, with westerly winds, and much mistiness in the atmosphere, such as would result from the partial mixing of two currents of unequal temperature. The mean temperature of the air during this time was 44°. On the 31st, however, the barometer fell rapidly, and by 10 A.M. on the following day was as low as 29·588 inches : heavy rain had fallen in the night from the south-west, and the great body of the air was then probably flowing from that quarter.

(175.) The greatest depressions of the barometer mostly occur during stormy and unsettled weather, or are coincident with a sudden transition from frost to thaw. Like great elevations, they generally, but not always, come on gradually. The day previous to the great depression of the 13th of January 1843, the weather had been fine, and the air perfectly calm, with sharp frost, and the wind N.W. But the state of the barometer, which was very low even then, indicated the near approach of some change, and in the night following the wind suddenly shifted to S.W., blowing at the same time very fresh, and attended by hard rain, and a rise of temperature. The barometer continued falling all the next morning, and on to 2 P.M. ; but for some hours previous to attaining its lowest point, both wind and rain partially subsided. The former, however, sprung up again with increased violence about sunset, and blew almost a hurricane till 7^h the next morning, at which time it was observed to be W.N.W. During the prevalence of this heavy gale, the barometer rose about seven-tenths.

(176.) As in the above instance, so in most others, I have noticed that, though a fall of the barometer, when considerable, is generally attended by much wind, the gale is not at its height till after the mercurial column has reached its greatest depression, and has begun to rise again. This appears due to the circumstance of its being the S.W. wind, which causes the barometer to fall, and the sudden shifting of the wind again to N.W., which causes the gale, at the same time making the barometer rise. Great falls of the barometer, however, sometimes occur without any wind at all.

(177.) During heavy gales, especially if the wind blow in gusts, the mercury in the barometer (at least in all good instruments having a perfect vacuum in the upper part of the tube) may be seen to heave up and down with considerable force. In one instance, on March 8th, 1827, during one of the strongest gales I ever remember, that blew with unabated violence for more than twelve hours, the oscillations of the barometer at Swaffham Bulbeck were most distinctly visible, the mercury rising and falling alternately through a space equalling ·02 inch, its surface first becoming convex, and then instantly changing to concave. Oscillations under one-hundredth of an inch, arising from this cause, are by no means unfrequent*.

(178.) As considerable elevations of the barometer *may* occur with a W. or S.W. wind, so in like manner the barometer *may be* very low with E. or N.E. winds. But as in the former instance there may be presumed to be an upper current from the N., which keeps the mercury high, so in the latter we may infer there is an upper current from the S.W., which keeps it low. Thus, for the first three days in November 1844, the mean height of the barometer was only 29·400 inches, with Easterly winds the whole time. During the last week in October, the wind had been nearly in the same quarter, with the barometer much higher and standing,

* Professor J. D. Forbes, during a hurricane which blew on the 7th of December, 1827, observed an oscillation of the mercurial column to the amount of at least ·03 inch. See *Edinb. Journ. of Sci.* vol. ix. p. 139.

until the last two days, above 30 inches. But it was evident that the upper current had been gradually working round to the S.W., through the N. and N.W., from the time the barometer began to fall, the lower, after a few days, following in its turn, and, till *it* also had revolved, causing the barometer to be very low, notwithstanding the vane still pointed E. In one instance I observed the barometer as low as 28·739 inches, the wind, *i. e.* the current next the earth, nevertheless being Easterly.

(179.) It is worth noting that the two greatest elevations of the barometer I ever observed occurred both at the hour of 10 P.M., which is coincident with the period at which the semi-diurnal oscillations take most effect. The two greatest depressions were at the hours of 1 A.M. and 2 P.M. respectively.

(180.) Great rises, as well as great falls, of the barometer are not generally confined to a few near-adjoining localities, but occur contemporaneously in very distant countries, extending over a wide region of the earth of several hundred miles. At the same time it is clear that, as the whole weight of the air remains the same, every elevation in one region must be accompanied by an equivalent depression somewhere else. Hence the fluctuations in the pressure give rise to what may be considered as atmospheric waves, more or less extensive in their operation, and which are transmitted from place to place at intervals of different lengths in different instances, though the intervals are the same, or nearly so, in the case of any particular wave. If in a given locality there occur an elevation of the atmosphere above our heads, it is followed sooner or later by a corresponding depression. It will be generally found that, after the barometer has stood high for a few weeks, accompanied by fine settled weather, an opposite state of things is brought about to restore the balance, the barometer remaining low for a considerable period, and the weather changeable.   It is not so much, however, in the ordinary, as in the more unusual and greater disturbances of the atmospheric equilibrium, that these waves can be distinctly traced.   In these latter the fluctuations of pressure, as shown by the barometer, are not only more marked, but seem to follow with more regularity.   It has been often noticed that a high extreme of the mercurial column is succeeded by a low extreme about a fortnight or so afterwards, though this may not invariably follow.   Dalton, many years ago, attempted an explanation of this circumstance, from which he was led to conclude that "we ought never to expect an extraordinary fall of the barometer, unless when an extraordinary rise has preceded, or at least a long and severe frost*."   More recently the subject of atmospheric waves has been ably investigated by Mr. Birt, in several communications made at different meetings of the British Association, and he has, in certain instances of great disturbance of the atmospheric pressure, traced the extent and progress, as well as the rate of progress, of these waves, by comparing registers of the barometer kept in various and distant localities, and noting down the several lines of contemporaneous elevation and depression, as also the actual heights of the mercurial column in each case.   It is remarkable in some of these instances in how similar a manner the barometer has been affected over a widely extended region, and how between the times of greatest elevation

* See his *Meteorological Observations and Essays*, p. 113.

and depression, marking the "crest" and "trough" of the wave respectively, there has been nearly the same interval at each of the places which the wave has successively passed over*.

(181.) It has been stated above (173) that great elevations as well as great depressions generally take place very gradually. This, however, is not always the case, and sometimes the rise of the mercury after heavy storms is very rapid. In one instance that I observed, it rose a whole inch in the course of a single night. When the barometer keeps rising and falling alternately in this rapid manner, it shows that there are great disturbances of the air over different regions of the globe, arising from the conflicts of opposite currents, and we have reason to expect very variable weather for a considerable time.

(182.) Sometimes, on the other hand, during certain states of weather, the barometer hardly varies one way or the other for days together. The most remarkable instance of this kind that ever came under my observation was one in which the mercury remained perfectly stationary at the height of 30·01 inches, from 6^h P.M. on the 23rd of August, to 9^h A.M. on the 27th, a period of 87 hours. The weather all this time was fine and settled, the wind steady in the E., but the air by no means calm, as might have been expected, and occasionally a stiff breeze blowing. Howard also has recorded an instance in which the barometer remained stationary for 60 hours†.

(183.) The mean annual range of the barometer at

* See *Reports of the Brit. Assoc.* (1844), p. 267; (1845), p. 112; (1846), p. 119; (1847), p. 351; (1848), p. 35.

† *Climate of London*, vol. iii. p. 217.

Swaffham Bulbeck is 1·890 inches.   The greatest annual range occurred in 1843, and amounted to 2·407
inches; the least, in 1832, being 1·397 inches.

(184.) From considering the results of the barometer
for each year, I pass on to those obtained for each
month in the year.

TABLE II.—*Showing the mean height of the barometer
at Swaffham Bulbeck, during each month of the year,
together with the greatest and average extremes, and
the greatest and average range.*

Month.	Mean Height.	Average of maxima.	Average of minima.	Difference or Mean Range.	Greatest Elevation in 19 years.	Greatest Depression in 19 years.	Difference or Full Range.
January ...	29·899	30·533	29·049	1·484	30·880	28·143	2·737
February .	29·874	30·433	29·088	1·345	30·866	28·648	2·218
March......	29·905	30·485	29·222	1·263	30·704	28·625	2·079
April	29·842	30·344	29·283	1·061	30·526	28·965	1·561
May	29·958	30·341	29·444	·897	30·572	29·056	1·516
June	29·902	30·289	29·458	·831	30·510	29·141	1·369
July	29·934	30·276	29·524	·752	30·400	29·205	1·195
August ...	29·872	30·267	29·404	·863	30·371	29·106	1·265
September.	29·899	30·291	29·399	·892	30·541	28·990	1·551
October ...	29·847	30·411	29·076	1·335	30·718	28·770	1·948
November .	29·803	30·408	29·059	1·349	30·587	28·510	2·077
December .	29·924	30·482	29·137	1·345	30·722	28·388	2·334

(185.) From the above Table, May would seem to
be the month in which the barometric mean rises the
highest, and November that in which it is lowest; and
it is worth noting that these are two *corresponding*
months, the first being the month before that of the
summer solstice, the last the month before that of the
winter one.   The difference between the two, or the
amount of variation of the mean in different months, is

·155 inch; and in no month does the mean height of
the barometer rise to 30 inches.

(186.) No regularity, however, can be traced in the
variations of the mean atmospheric pressure from month
to month, perhaps in consequence of the results being
more correct in some months than in others; but if we
take each of the *seasons* separately, there would seem to
be a slight increase of the pressure from autumn, in
which it is lowest, to summer, in which it is highest.

inches.

The mean height of the barometer for Spring   is 29·901.

,,   ,,   ,,   Summer is 29·902.

,·   ,,   ,,   Autumn is 29·849.

,,   ,,   ,,   Winter is 29·899.

The difference between winter, spring, and summer
is scarcely appreciable, but between summer and autumn it amounts to ·053 inch.   Howard found, on a
careful examination of daily observations carried on for
a period of ten years, " that the winter quarter, beginning with the solstice, had gained upon the autumnal
quarter, beginning with the equinox, ·021 inch,—that
the spring quarter had gained upon the winter ·030
inch,—that the summer had gained upon the spring
·045 inch.   But in the *autumn*, the whole increase
went off again, the barometer averaging in this quarter
·096 (or nearly a tenth of an inch) below the summer.
The column, then (he adds), stands highest in the latter
part of summer, and lowest at the beginning of winter *."
This difference in the barometer between these two seasons, Howard attributes, and perhaps rightly, to the
increase of vapour in the air during the summer, in
consequence of the heat.   This, of course, added to the

* *Lectures on Meteorology*, 2nd Edit. p. 77.

weight of the air itself, increases the pressure of the whole atmosphere.   In the autumn, however, the great excess of vapour is deposited in rain, and the pressure diminishes.

(187.) The month in the above Table in which the mean height of the barometer approaches most nearly to the mean annual height is February, but this result, in the present instance, may be merely accidental, and consequently unimportant.

(188.) It has been already stated (172) that the greatest elevations of the barometer, as well as the greatest depressions, all occur during the winter months ; and it will be seen that this is true, as well in respect of the *average* maximum and minimum of each month, as of the two widest extremes during the whole period for which the observations were continued.   The elevations are most considerable in January, from which month they fall off with very little irregularity to the month of August, in which they are lowest; and the difference between the average maxima of these two months is rather more than a quarter of an inch.   From August to January they again gradually rise.   The depressions, too, are most considerable in January, from whence they gradually decrease to July ; and the difference between the average minima of these two months is nearly equal to half an inch.   It necessarily follows, that both the *mean* and the *full* monthly range will be also greatest in winter, and least in summer, as is evident at a glance of the contents of the above Table.

(189.) *Height of the Barometer in connexion with the direction of the Wind.*—The height of the barometer varies with the direction of the wind.   It has been already stated that Northerly and Easterly winds have a tendency to raise the mercurial column (163), while Southerly and Westerly winds tend to depress it (162). But if we take the average of a large number of observations of the barometer, under the influence of each of the winds in succession, confining ourselves for this purpose to the eight chief points of the compass, we shall find a regular fall or rise of the mean result each way, according as we pass from N.E. to S.W., or, contrariwise, from S.W. to N.E. Even the observations of a few months will serve to show an approximation to this regularity, though a much larger number are required to determine the exact mean height of the barometer corresponding to each particular wind. There are no doubt some anomalies; and many places, at least in other countries, if not in our own, in which greater or less deviations from this rule will be found to occur. And where the rule may generally hold good, the particular mean for each wind will doubtless be somewhat different in different localities. But observations, upon which to ground any comparison of this kind, do not appear to have been very frequently made in this country, though the subject has received more attention abroad. The results of some of those made on the Continent have been collected by Kämtz, who has given the mean pressure of the barometer at fifteen places in Europe, with the eight principal winds. One of these places is London, the results for which are given below, in order to compare them with my own, with which they do not exactly agree. The difference between the two may be due in part to my own observations not having been sufficiently numerous, in part also, perhaps, to difference of locality.

				London. inches.	Swaffham Bulbeck. inches.
With the wind N. barometer averages				29·89	30·041
„	N.E.	„	„	29·95	30·075
„	E.	„	„	29·88	29·875
„	S.E.	„	„	29·78	29·750
„	S.	„	„	29·69	29·744
„	S.W.	„	„	29·73	29·727
„	W.	„	„	29·81	29·925
„	N.W.	„	„	29·84	29·946

Kämtz infers from all the results he has obtained, that the barometer stands highest when the wind blows from between the E. and the N.; and lowest when it comes from a point comprised between the S. and the W. This agrees with the above as regards the maximum height, which, both at London and Swaffham Bulbeck, occurs with the wind N.E.; and the circumstance of the minimum height occurring apparently with a S. wind at the former place and with a S.W. wind at the latter, indicates no very great discrepancy, if the real minimum lies, as it probably does, somewhere between these two points. The differences between the several heights in the above Table taken consecutively, are not uniform in the two cases, but many local influences, independent of other causes, may tend to bring this about. It is of great importance, however, for judging of the weather, to know, in a general way, what the mean height of the barometer for each wind is, as a height which would indicate a tolerably settled state of the weather with the wind S. or S.W., would not be an equal security against rain, if the wind were in a quarter directly opposite. To afford such indication, *the barometer should be above the mean corresponding to the particular wind blowing at the time.*

(190.) With the knowledge of the above rule, it is extremely interesting to watch the variations of the barometer, indicating, as they do, the different shiftings of the wind in the upper regions, often before there is any appearance of change in the vane below. Thus if the wind be S. or S.W., and the barometer stand about the mean height corresponding to those winds, and we shortly afterwards observe a rise in the mercury, we may be pretty sure that the upper current is inclining more or less to the N.W. Or if, being S.W., it pass through the S. to the S.E., we may expect that it is not about to remain there, but to proceed on to E. or N.E. In like manner, when the wind is E. or N.E., a slight fall in the barometer below the mean for those winds, indicates a tendency towards the S. or N.; and if the fall continue, the wind will eventually work round and settle in the S.W. If the barometer be very low when the wind is Easterly, or very high when in the S.W., as it sometimes is for days together, we may judge that there is an upper current all the while setting steadily in from the opposite direction in each case; and so long as this continues, the weather will probably be very wet in the former instance, and very fine in the latter. A gradual rise of the barometer in the first case, and a gradual fall in the second, speak to the upper current working slowly back to fall into the same direction as the lower, which latter may not have varied during the whole time. Again, as the movement of the barometer tells us of the movement of the upper current, while the lower remains stationary, so, on the other hand, the barometer remaining stationary, or varying but very little, tells us that the great body of the atmosphere up above is either at rest, or moving steadily in one direction, notwithstanding the current next the earth, as indicated by the vane, may oscillate greatly, or even go completely round the compass.   Hence, too, we may generally infer that, if the wind shift from the S.W. to the N. and N.E., or, contrariwise, from N.E. to S.W., without affecting the barometer to any considerable extent, it is a mere local transfer of the lower current, which, although it may for a time cause the sky to become overcast, or bring about some little alteration of weather, will not last long, the vane soon returning to its former direction.

(191.) Another circumstance worthy of observation is that, as the barometer stands highest with N.E. winds and lowest with S.W., and as these are the two winds which have a tendency to prevail over the rest, it is when the wind is in one or other of these two quarters that the barometer is steadiest, being generally in a fluctuating state at other times.   The wind, if we may so speak, is always more or less striving after one of these two points ; when N., it is striving to get on to N.E., consequently the barometer almost always rises on those days on which it blows from the quarter first mentioned.   When the wind reaches the N.E., the pressure is at its maximum, and the barometer varies but little during the day.   In like manner, when the wind has quitted the N.E., and is about to pass to S.W., through E., S.E. and S., which is the direction it generally takes, it may remain, for a short time, during its passage, in each of the last three quarters in succession, but the barometer is in a falling state, and continues so, until the wind has arrived at the S.W., when the pressure attains its minimum, and the mercury is again comparatively steady. It must be remembered, however, that these results will

be often masked by the diurnal oscillations of the barometer, which tend to make the mercurial column lower
in the middle of the day than in the morning and
evening, and which, being due to the calorific influence
of the sun, are quite independent of any variations of
the instrument from the above causes. It is only when
we compare the height of the barometer at different
times of the day, under the influence of each wind, in a
large number of instances, that we can perceive its tendency to remain stationary for a time with certain winds,
and to rise or fall with others.

(192.) *State of the Barometer during rain.*—We are
led to infer from what has been said above (189), that
as the barometer must be above the mean corresponding to the particular wind blowing at any time to ensure
fine weather, so it must be below that mean to bring
about continuous wet. And this in a general way is
true. Rain may fall in the form of light showers, or even
heavily for a short time, under any conditions of the
barometer. In one exceptional instance, on the 28th
of May 1849, it continued falling for several hours,
with a rising barometer, never lower during any part of
the day than 30·20 inches. I have even known a case,
in which there was *some* rain, the barometer standing
all the while as high as 30·62 inches. In this last instance, as in many similar instances of rain with a high
barometer, the rain was of a light mizzling character,
and appeared due to the great body of the atmosphere working *very gradually* round from S.W. to N.E.,
whilst the air was in an extremely humid state. Ordinarily, however, these anomalies are mere local occurrences, of too limited extent to exercise any material
influence upon the mercury, and consequently not last-

ing in their effects. As a general rule, the barometer
will be found standing at or near a certain given height,
during what may be called " regular wet days"; and also
gradually falling to that point, previous to wet or
changeable weather of many days' continuance setting in.

(193.) Before proceeding to point out what that
height is, it may be observed that most of the rainy
weather which occurs in this climate is either the result
of a S.W. wind, loaded with vapours, blowing next the
earth, or of a S.W. current overlying one from the E.
or N.E. In either case the barometer would stand
lower than the general mean. In the latter instance,
cold and wet occur together, giving rise to very disagreeable weather: the barometer, however, would stand
rather higher in this than in the former case, in consequence of the colder current below being heavier than
the one above it. To whichever state of things the rain
is due, many days often elapse before the air at the
place of observation becomes so thoroughly saturated,
or the two currents so thoroughly mixed, as to cause the
rain actually to descend. During this interval the sky
keeps getting more and more overcast, and the barometer keeps sinking. Sometimes the latter will continue to fall for a whole week or more, previous to any
rain taking place, though followed by wet in the end;
and when, at last, the rain does come, it will often be
found that the barometer begins to rise again, in consequence of the S.W. wind being gradually opposed
and borne back by a current from the opposite quarter.
So long as the S.W. wind prevails, the barometer will
fall; but the rain may not ensue till the N.E. wind
mixes with it, causing, by its lower temperature, both a

precipitation of the vapour which the S.W. wind had brought up, and a rise of the mercurial column.

(194.) The following Table will now show what, in general, the actual height of the barometer is during settled wet in this country, or at least, in that particular neighbourhood in which my observations were made. The results therein given have been obtained by putting together a very large number of instances of rainy days, with the corresponding winds, and state of the barometer under each wind, and then taking the means severally.

The mean height of the barometer during rainy weather, with the wind—

		inches.				inch.
N.	averages	29·661,	being below the average mean			·380
N.E.	„	29·787	„	„	„	·288
E.	„	29·682	„	„	„	·193
S.E.	„	29·621	„	„	„	·129
S.	„	29·473	„	„	„	·271
S.W.	„	29·528	„	„	„	·199
W.	„	29·603	„	„	„	·322
N.W.	„	29·689	„	„	„	·257
	Mean....	29·630		Mean....		·255

The mean of the whole, or the mean height of the barometer during wet weather, without reference to any particular wind, is seen to be 29·63 inches (coinciding nearly with the mean of the average yearly extremes), while the mean fall below the general average is ·255 inch. But as by far the larger quantity of the rain that falls in this country comes from the S. and S.W., the average state of the barometer during wet will be more fairly estimated by taking its mean height under the prevalence of those winds only, which, when the two are combined, gives a result of 29·50 inches, being

·235 inch below the general mean for the same two winds. This corresponds with the point at which the word *Change* is marked on ordinary barometers, while *Rain* is set at 29 inches, or half an inch lower. Hence it is clear that these words, as at present fixed, must be understood with great latitude, if they have any meaning at all. In fact, settled wet, as above seen, more often comes on at that height of the barometer at present marked *Change*, while the first indications of change, or, as it were, the turning-point from dry to wet or wet to dry, might more properly be fixed two- or three-tenths of an inch higher. And even thus determined, the points in question must be considered as having reference only to wet coming from the S. or S.W., and to localities not much raised above the level of the sea. With any other winds the same quantity of rain may be attended by a higher state of the barometer, as indicated in the above Table: also, if the spot in which the observer is situate is much elevated, due allowance must be made for such elevation; the higher the place is, the lower of course being the barometer, in respect of its average height, as well during dry weather as wet.

(195.) Dalton has made an observation, which agrees with what I have often noticed myself, and which, if really correct, shows still more forcibly the error of setting the word *Rain* against a height of the barometer answering to 29 inches, and *Much Rain* at 28·5 inches. He states, as the result of a careful examination of a register kept at Kendal for several years, that, though we may expect rain in proportion as the barometer falls below its mean annual height, yet this only holds true till it descends to a certain point, below which, if the mercury continue to fall, the rain ordinarily diminishes. By taking the mean height of the barometer on those days on which the greatest quantity of rain had fallen, he found that the heaviest rains might be expected when the barometer is about 29·47, or in round numbers $29\frac{1}{2}$ inches, which is a little *above* the mean of two great extremes observed by him in January 1789, or 29·44 inches. Taking again the average quantity of rain on certain other days when the mean state of the barometer was below 29 inches, he found that the fall was much less than in the former instance, though there was very little really fair weather. Hence he concludes "that when the barometer is very low, the probability of its being fair is much smaller than at other times; but that, on the other hand, the probability of very much rain, in 24 hours, is not so great as at other times." This is quite at variance with the common notion, that the lower the barometer gets, the more rain there will be; though if we consider the fall of the barometer as measuring the probability of tempestuous weather, in respect of *wind*, rather than of any *large amount of rain*, our expectations will generally be right.

(196.) Dalton considers this circumstance of the greatest rains occurring in connexion with a barometric height of about $29\frac{1}{2}$ inches, or the mean of the two average extremes, as apparently consistent with the theories of the barometer and rain; and he explains it in the following manner. He says, " When the barometer is above the mean high extreme for the season of the year, the air must, relatively speaking, be extremely *dry* or *cold*, or both, for the season; if it be extremely dry, it is in a state for imbibing vapour; and if it be extremely cold, no further degree of cold can then be expected, and therefore in neither case can there be any considerable precipitation : on the contrary, when the barometer is very low for the season, the air must relatively be extremely *warm* or extremely *moist*, or both ; if it be extremely warm, it is in a similar state to dry air for imbibing vapour, and if it be extremely moist, there must be a degree of cold introduced to precipitate the vapour, which cold, at the same time, raises the barometer.   From which it follows, that no very heavy and continued rains can be expected to happen whilst the barometer actually remains about the low extreme, but they must rather be the consequence of a junction or meeting of extremes, which at the same time effects a mean state of the barometer*."

It will be at once seen how exactly the height of the barometer during the heaviest rains, as set by Dalton, agrees with the height determined by myself, and stated in the remarks which follow the Table above given.

* *Meteorological Observations and Essays*, p. 151.

CHAPTER IV.

AQUEOUS PHENOMENA OF THE ATMOSPHERE.

EVAPORATION AND THE DEW-POINT.

ALLUSION has been already made to the aqueous vapour which is constantly mixed, though in ever-varying proportions, with the atmosphere of air above our heads. It has been further stated that the source of this vapour is the water evaporated from the surface of the earth (161). The phenomena dependent upon the quantity contained in the air at different times, such as dew, mists, clouds, rain, &c., offer some of the most interesting subjects for consideration in the science of Meteorology.

(197.) Not only is the amount of vapour in the atmosphere very variable, but, as is well known, some districts and countries are always much more humid than others. Such is the case with that part of Cambridgeshire, bordering upon the fens, in which my own observations have been made. It is a great mistake, however, to suppose, as some do, that this is owing to any excess of rain falling at such places, between which and the humidity of a climate there is no necessary connexion. It will be seen from the tables hereafter given, that as little rain falls at Swaffham Bulbeck as, perhaps, in any other part of England: the dampness of that locality, therefore, must be due to some other cause. It arises in fact from its low marshy character, and from the nature of the soil. The latter being at bottom a stiff clay, prevents the rain from soaking-in to any considerable depth : the wet therefore accumulates near the surface, keeping the lower strata of the superincumbent atmosphere always in a comparatively moist state.

(198.) Likewise, it is not the *actual*, but the *relative* quantity of vapour contained in the air at any time that determines its humidity. It is the quantity present in relation to the quantity required to saturate it at the existing temperature. Hence in winter, when the temperature is low, the air generally feels damp, notwithstanding a much less amount of vapour actually exists in it at that season than in summer, when from the higher temperature it feels comparatively dry.

(199.) Evaporation, by which the supply of vapour is kept up, would seem to be always going on, though at a rate varying according to the temperature, the amount of vapour already existing in the air, and the weight of the superincumbent atmosphere*. Howard observes that " it is not always suspended even during rain, though the rate is much less on those days in which rain falls, and it is liable to a rapid increase immediately afterwards†." The circumstances which

---

* It is not, however, *by its weight*, as Dalton has shown, that the atmosphere obstructs the diffusion of vapour, for this would effectually prevent any vapour from rising under the temperature of 212°; but it is by the *vis inertiæ* of the particles of the air. Dalton describes the obstruction as " similar to that which a stream of water meets with in descending amongst pebbles."

† *Climate of London,* vol. i. p. 85.

most favour evaporation are heat, dry air, and a diminished pressure on the evaporating surface. When, on the contrary, the temperature is low, the air already saturated with moisture, as during fog, and the barometer high, a state of weather which often occurs in winter, evaporation is reduced to a very low point, though perhaps never altogether arrested.

(200.) It is remarkable that evaporation still goes on when water is frozen, the same as when it is liquid[*]. Even "the most intense cold is insufficient of itself to put a stop to it[†]." This circumstance often strikes persons with astonishment, who witness it in its effects, without being aware of the true cause. They see a fall of snow gradually waste—if light, wholly disappear, or a block of ice sensibly diminish, during the continuance of a frost, especially if the wind blow tolerably fresh from some point towards the N., without the least signs of liquefaction on the surface. They wonder what is become of it. " Sometimes, also, in deeper snows, the surface becomes curiously grooved and channeled by the wind acting unequally on it and thus promoting unequally the evaporating power. This phenomenon is best observed around the trunks of trees and near the interstices of palings, or wherever a stream of air acquires an increased force in a particular direction[‡]."

* Dr. Prout observes that not only "is evaporation constantly going on from snow and ice, but there is every reason to believe that the quantity of vapour thus formed from snow and ice is precisely equal to what would be evaporated from water itself, provided water could exist as a fluid below the temperature at which it is congealed."—*Bridgewater Treatise*, p. 276.

† Howard.   ‡ *Encycl. Metrop.*, art. " Meteorology."

(201.) A high temperature being one of the conditions necessary for a free evaporation, this last varies, as might be expected, with the season. But the rate of its increase and decrease is not always exactly proportional to the increase and decrease of the temperature, in consequence of other causes interfering to influence the amount. Thus the evaporation is greatest in summer and least in winter. But during the spring it exceeds what takes place in autumn, notwithstanding the higher mean temperature of the latter season as compared with the former. This is evidently due to the dry Northerly and Easterly winds which usually prevail during the spring months, and which allow of a much larger amount of vapour being taken up.

(202.) Howard states "the common rate of evaporation per day from a surface of water exposed to a free air" to be, "in winter, from a tenth of an inch down to a hundredth; in summer, from two to three tenths*." The greatest evaporation in one day he ever observed was 0·39 inch.

(203.) The total annual amount of evaporation was found by Howard one year nearly to equal the annual average depth of rain about London; but it is not likely that it ordinarily comes up to this mark, since if this were everywhere the case, as Dalton has observed, there could be no rivers†.

(204.) It is the sea of vapour, mixed up with the air in greater or less quantity, which, quite as much as temperature, assists in impressing upon climate its peculiar character, more especially when that climate is, like our own, insular. Surrounded on all sides by the ocean, constant supplies of vapour are taken up by eva-

* *Lectures on Meteorology*, p. 84.   † *Met. Essays*, p. 137.

poration, which, when condensed into clouds, serves to moderate both the heat of summer and the cold of winter in England, compared with other countries in the same latitude in the interior of continents. The vapour, likewise, according as it is in excess or not, greatly determines the effect which each particular season exerts upon the human constitution. Combined with Westerly winds and a moderate temperature, neither hot enough to oppress nor cold enough to chill, it is the moisture in the air which renders autumn in general so peculiarly agreeable to the feelings of man. But as the year declines, and winter draws on, the moisture becomes superabundant, a precipitation in the form either of mist or rain takes place, while the temperature falls; and unless the cold be sufficient to cause frost, the excess of humidity communicates to the air that *raw* feel, as it is termed, which is no less unpleasant to the feelings than prejudicial to the health. It is at such times that bronchial diseases and rheumatic affections prevail, especially if the temperature continue for any length of time just above freezing-point, without actually descending to it*

* This was very much the character of the weather at Swaffham Bulbeck, during the second and third weeks in January 1837, when the influenza was so generally prevalent throughout the kingdom. After a fortnight of severe frost, accompanied by a heavy fall of snow, a thaw took place on the 6th of that month, but the rise of temperature, for several days, was inconsiderable, and dull monotonous weather ensued, with much mist and little sun, the thermometer neither descending (or only just descending) to the freezing-point, nor yet rising much above it. All this time the air was in an extremely humid state, from the partially melted snow, and sleety rains occasionally. The barometer was for the most part high: the wind variable. For some

(205.) Again, we all know the effect upon the human frame of the sharp cutting Easterly winds of spring; but the chilling sensations which they produce are not due mainly to their temperature (for we often have a temperature quite as low at other periods of the year, without experiencing the same discomfort from it), but to their dryness: in consequence of this, they readily take up moisture wherever they can find it, the human body being made to yield its share, whereby the skin becomes parched and chapped, as well as colder than it would otherwise be rendered, owing to the evaporation that takes place from its surface.

(206.) When, again, the moisture in the air is in great excess, combined with a high temperature and Southerly winds, as frequently happens in the hot months of summer, the skin becomes covered with perspiration, while the evaporation is checked. This causes a feeling of lassitude, which prostrates the strength and energies of many persons, and almost forbids any exertion. It is in this sultry state of the air that we often expect thunder; and though the prognostic may sometimes fail, no condition of the atmosphere is a more general precursor of thunder-storms in hot weather, than a rapid increase of the vapour contained in it, as shown by a greatly diminished evaporation from wet or damp surfaces.

(207.) The aqueous vapour in the air thus playing so important a part, it is very desirable to have some means of measuring its amount from time to time. On the nature and structure of the instruments used for this purpose it is not my intention to dwell. There

account of this epidemic, and the weather accompanying it, see Dr. Holland's *Medical Notes and Reflections*, p. 186.

can be little doubt that the most convenient and most readily observed is the wet- and dry-bulb thermometer. But at the time that I commenced my observations, this instrument had not been constructed in the same perfect and careful manner in which it is now, nor the same accurate tables published, which have been since*, for estimating, from the indications it affords, the true dew-point, and other data connected with the humidity of the air. I was led, therefore, to employ Daniell's hygrometer, with which I made observations nearly every day for a few years at 10 A.M. This instrument, however, though valuable in many ways, is but ill-adapted for constant use, owing to several practical difficulties in the management of it. The large quantity of æther which it requires, and the difficulty of obtaining the æther good, together with the impossibility of getting a dew-point, even with the best and strongest æther, in very dry weather, combine to make this kind of hygrometer both troublesome and objectionable. To which may be added, that it requires a well-practised eye to note the exact moment at which the ring of condensed vapour forms on the darkened bulb, and, at the same time, the height of the thermometer within. It has been sometimes recommended, with the view of obviating this last difficulty, to observe, not merely the first appearance, but also the moment of disappearance, of the ring of vapour, and to take the mean of the two temperatures as a more correct indication of the true dew-point. But it will be seen from the following experiments which I made, that the moment of disappearance is not always the same, being sometimes above and sometimes below the temperature

* See especially Glaisher's *Hygrometrical Tables*, 1847.

at which the dew first forms on the bulb, and dependent, it would seem, upon the amount of dew formed,
and the quantity of æther used. The experiments were
made on a fine summer day at 2 P.M., in the middle of
an exposed grass-plot, the sun being partially concealed
by clouds, and the temperature of the air being 74°.
They were repeated five times, allowing only a sufficient
interval between to enable the instrument after being
used to return to its original state. The temperature
at which the ring of vapour first showed itself was
found to be 53° in all the experiments, but that at
which it finally disappeared was different in four out of
the five, as seen below :—

(Ex. 1.) Not much deposition of dew: ring of vapour disappeared at 51°.

(Ex. 2.) Rather more deposition: ring of vapour disappeared
at 52°·5.

(Ex. 3.) Not much deposition: ring of vapour disappeared at
52°. *N.B.* The thermometer in the bulb remained at 52° several seconds free from vapour, and then gradually rose again,
without any further deposition.

(Ex. 4.) The circumstances exactly the same as in the last
experiment: ring of vapour disappearing at 52°.

(Ex. 5.) Rather more æther poured on, which caused a more
copious deposition of vapour: ring did not wholly disappear
till the thermometer had risen to 55°·5.

On another occasion of repeating this experiment,
after a dew-point had been obtained at the temperature
of 55°, that of the air being 67°, I found that by continuing the application of æther until the thermometer
in the bulb was depressed to 45°, being then 22° below
the temperature of the air, the increased deposition of
vapour so produced did not disappear till the thermometer had again risen to 60°.

I

(208.) These experiments, in connexion with what was previously observed, show the caution required in using this instrument. When the dew-point, or the temperature at which the existing vapour can no longer maintain its aëriform state, has been correctly ascertained, we learn thereby the hygrometric condition of the atmosphere. The greater the difference between that temperature and the temperature of the air, the drier the air is, and the less chance of rain. It must be remembered, however, that this indication is quite independent of the *absolute* quantity of vapour existing in the atmosphere, which last, as before stated (198), varying with the season, may be more considerable at one time, when the air is very dry, than at another, when it is very moist. The general character of any climate in respect of humidity, would be determined by the mean dew-point as compared with the mean temperature. I much regret that my own observations with the hygrometer at Swaffham Bulbeck have not been sufficient to afford all the results that might be desired on this subject. The mean annual dew-point I believe to be $45°·5$, or rather more; but as in certain months of the year the observations were not conducted with the same regularity with which they were in others, I am not able to speak with confidence as to the mean dew-point of each month and each season in particular. The mean temperature of Swaffham Bulbeck being set at $49°$, the difference between that and the dew-point, equalling $3°·5$, will be the mean dryness for that locality. This is $1\frac{1}{2}$ degree less than at London, according to Daniell, who gives the mean temperature there as $49°·5$, and the mean dew-point as $44°·5$, the difference between which is $5°$. This circumstance is indicative of the greater humidity of the climate in that part of Cambridgeshire than in the metropolitan district.

Though unable to fix with precision the mean dew-point of each separate month at Swaffham Bulbeck, the collected observations I have made in reference to it lead to the inference that the dew-point is lowest in January, and highest in August, gradually advancing from the minimum to the maximum, and then gradually declining from the maximum to the minimum, though in each case by unequal steps in the different months. This is much in accordance with what takes place in the neighbourhood of London; and it may not be without interest to give the mean temperature and the mean dew-point of each month in that neighbourhood, as determined by Daniell, together with the mean dryness, which, in its *relative* variations from month to month, does not perhaps differ very materially from what it is in Cambridgeshire. The data upon which the following Table is constructed are to be found in his *Essay on the Climate of London**.

Month.	Mean Temperature.	Mean Dewpoint.	Difference, or Mean Dryness.
January	36·1	34·3	1·8
February	38	34·9	3·1
March	43·9	39	4·9
April	49·9	43·5	6·4
May	54	46·1	7·9
June	58·7	50·7	8
July	61	54·5	6·5
August	61·6	55·3	6·3
September	57·8	52·3	5·5
October	48·9	44·8	4·1
November	42·9	40·5	2·4
December	39·3	37·6	1·7

* *Meteorological Essays*, pp. 263-305.

I 2

(209.) The relative advances of the temperature and
the dew-point during the first half of the year are
brought distinctly into view in the above Table; and
Daniell has well observed how admirably Nature's order
is adapted for securing a right state of the earth just at
that period of the spring, in which the operations of
the farmer are so actively called forth.   March and
April are two of the most important months to him;
and the preparing of the soil, and the subsequent sowing of the seed, upon which his harvest is to depend,
could hardly be conducted with any success under other
conditions than those which ordinarily exist at that particular season.   It is just then, when the superfluous
moisture is being more and more exhaled each day,
that a dry state of the earth is most needed.   During
the two winter months of December and January, the
atmosphere and ground are alike nearly saturated with
moisture, the quantity of rain greatly exceeding the
amount of evaporation.   In February, the wet of winter
begins to dry up; the temperature slightly rising, while
the dew-point continues much as in the preceding
month, and the quantity of rain and the amount of
evaporation are brought nearly to an equality.   But
after this a great step is somewhat suddenly made
towards lessening the humidity: the mean dryness, or
the difference between the temperature and the dew-
point, advances in March by nearly 2 degrees, and
keeps continually on the increase till the month of
June, in which it attains its maximum.

(210.) While the dry state of the air in spring thus
favours the operations of husbandry, it leads us to see
the impropriety of choosing that season for planting,
which, though sometimes deferred till after winter,

should be always done in autumn, when the humidity
is in excess.   It is scarcely less important, after shrubs
and trees have been transplanted, that they should have
a humid atmosphere about their branches, than a humid
soil about their roots.   Now both these advantages are
secured to them when moved towards the end of the
year; while there is then sufficient time to allow of
their getting well-rooted before being exposed to the
dry cutting winds of March and April.   And this will
further appear, when we take into account the intensity
of the sun's rays in spring, owing to the very circumstance of there being so little opake vapour to intercept
them in their passage to the earth.   Daniell has noticed
an extremely dry state of the air, which sometimes prevails for a few hours of the day, in the months of April,
May, and June, especially under south walls, which
become very much heated from the above cause, while
the dew-point is yet comparatively very low.   He mentions having seen in the month of May the thermometer in the sun at 101°, while the dew-point was only
34°*.   It may be imagined what must be the parching
effect of such an extreme state of dryness on vegetation
in general, and how prejudicial it must be, in particular,
to the tender shoots of fresh transplanted shrubs, which
have not sufficient vigour to resist its influence.

(211.) There is a close analogy, in the *diurnal* progress of the dew-point, and the variations in the amount
of vapour in the atmosphere, to what takes place in the
annual.   In this latter the course is as follows :—The
dew-point is lowest at the coldest period of the year :
the quantity of vapour is also least then ; though, from
the low state of the temperature, the humidity of the

* *Meteorological Essays*, vol. ii. p. 516.

air is greatest. As the year advances, the dew-point
and the temperature both rise; the quantity of vapour
also increases, owing to evaporation becoming more
active from the increased heat; nevertheless, from the
temperature rising so much faster than the dew-point,
the air keeps getting continually drier till June, in
which month the advances made by the temperature
and the dew-point are nearly equal. In July, the
quantity of vapour and the temperature both attain
their maximum; but, from the dew-point being likewise
very high, as well in this month as in August, the dryness, or relative humidity, begins to fall off, and an in-
verse order of things gradually takes place, until they
are brought round, at the end of the year, to the point
from which they set out.

(212.) It is much the same in the *diurnal* progress.
The dew-point and the quantity of vapour are both at
their *minimum* in the coldest period of the night, probably a little before sunrise, but the humidity is then
at its *maximum*. As the sun rises, and gets continually
higher, the air, with the increase of temperature, is
continually receiving fresh accessions of vapour, while
the relative humidity, or the difference between the
dew-point and the temperature, keeps diminishing.
This state of things goes on until the latter has attained
its maximum at some period of the afternoon, dependent upon the season. It is then to a certain extent
reversed, as in the former case. There is a difference,
however, in this respect in winter and summer. In
the former season, it is much as described above. In
summer, the quantity of vapour has a *double maximum*:
the first occurs before mid-day, sooner or later according to the particular month; after which "the absolute

quantity of vapour diminishes until the time of the highest temperature of the day, without, however, attaining a minimum so low as that of the morning." This is not due to any check given to the evaporation, for this last is most active when the temperature is highest; but to the ascending current drawing off to the upper parts of the atmosphere a great portion of the vapour raised from the soil, causing a diminution at the surface, where, at an earlier hour, it had accumulated, by virtue of the resistance of the air. " Towards evening, when the temperature begins to fall, the ascending current diminishes in force, or even ceases altogether; then, not only does the vapour accumulate in the lower parts, but it even descends from the higher regions;" giving rise to a *second maximum*, which, however, is not sustained, because, during the night, a portion of the vapour is precipitated in the form of dew*.

(213.) A close connexion, amounting almost to a complete coincidence, has been found to exist between the mean dew-point and the mean minimum temperature, the knowledge of which may be often of service to persons possessed of a self-registering thermometer, but without an hygrometer, in respect of estimating the relative quantity of vapour in the air, and the chance of its precipitation in the form of rain. This connexion was first pointed out by Dr. Anderson in a memoir published in 1824†, in which he has shown that the

* Much of the above account of the *diurnal* variations in the quantity of vapour is taken from Kämtz. It is probably applicable to most places in our own latitudes situate inland, and at no great elevation above the sea. See his *Meteorology* (by Walker) pp. 85 and 88.

† "On the Influence of the Hygrometric State of the Atmo-

minimum temperature of the night is, in fact, dependent upon, and regulated by, the constituent temperature
of the aqueous atmosphere. The thermometer ordinarily falls at night until it has reached that degree of
cold at which the existing vapour can no longer be
maintained in the aëriform state. The air, indeed,
may be so dry in certain states of weather, that the
temperature may cease falling before it has quite sunk
to that point; and it is under such circumstances that
we have those brilliant starry nights (during frost especially, which is always most severe when the air is thus
dry) that are the admiration of the beholder. But
more frequently, at an earlier or a later hour, the dew-

sphere upon the Minimum Temperature of the Night,"—in the
Edinb. Phil. Journ. vol. xi. p. 161.

In a subsequent communication to the British Association,
Dr. Anderson has pointed out "the *exact* coincidence which
holds between the dew-point and the minimum nocturnal temperature."—*Rep. Brit. Assoc.* 1840 (Trans. of Sects.), p. 41.

This subject has been since investigated by others. See,
especially, Dr. Martin's work on *The Undercliff* (p. 86), in which
he has given several comparative Tables illustrative of it. He
expresses his belief "that the minimum temperature of the
twenty-four hours, and the mean dew-point derived from *two*
observations taken at appropriate hours, approximate very
closely." At the same time he observes, "that under certain
conditions of the atmosphere the dew-point is many degrees
higher than the minimum temperature; and this especially
happens when severe weather is about to break up. The reverse
is frequently the case during severe cold, the hygrometer at
such periods indicating a *lower* dew-point than the minimum
temperature. The direction of the wind also tends to vary the
relation between the dew-point and the lowest temperature,
the hygrometer showing a lower degree in northerly and north-
easterly winds."

point is attained, when immediately a deposition of some
portion of the vapour takes place, causing an extrication
of latent heat, which for a time checks all further
descent of the temperature.   This agreement between
the minimum temperature and the dew-point, Dr. Anderson observes, "is most remarkable from the end of
July to the end of December, the temperature of the
year being then on the decline, and rendering the relative humidity greater than during the other half of the
year; but at no season is the deviation so great as not
to indicate a mutual connexion between them."

(214.) If, then, we have a thermometer that indicates
what the lowest temperature has been during the night,
it indicates with tolerable precision what the dew-point
has been also; and the difference between that minimum
temperature and the temperature on the following morning will be no bad guide to the character of the weather
during the day.   Now I find, on an average, as stated
in a former part of this work (38), a difference, in fine
settled weather, of about 20° between the minimum
temperature at night and the maximum of the ensuing
day.   These two extremes approach nearer to each
other, or there is less difference between the temperatures of night and day, in proportion as the relative
humidity of the air increases.   During wet, if the wet
continue for the twenty-four hours, they are often nearly
coincident.   If, then, we find the thermometer, say at 9
A.M., only 2 or 3 degrees above what it has been in the
night, and especially *if the minimum itself be high*, unless the wind change, causing the moisture to be again
taken up, there is considerable chance of rain.   If,
on the contrary, the thermometer at 9 A.M. be 10°, or
nearly 10°, higher than the minimum of the preceding

night, it will probably rise as much higher again during the day, and the weather continue fine.

(215.) But the knowledge of this coincidence between the minimum temperature and the dew-point may be turned to another account. In like manner as we infer the quantity of moisture in the air from ascertaining the minimum temperature, we may, if we have an hygrometer, infer what the minimum temperature will be any night by ascertaining the dew-point the preceding evening. I need scarcely mention the importance, to gardeners especially, of being able thus to determine beforehand how cold the night is likely to be,— or allude to the many instances in which, by taking timely precautions against frost, they might save their more tender fruits and vegetables from being killed.

(216.) It must be remembered, in all cases of making observations with the hygrometer, that this instrument simply indicates the degree of humidity of the air at the spot where the observer is situate. A very different state of things may prevail at a certain elevation above him, which renders it necessary that other considerations be taken into the account, before judging of the weather from the hygrometer alone. What the exact law is, which regulates the variation of the quantity of vapour in the air at different heights, is perhaps doubtful. It is conceived by most meteorologists, that the upper regions of the atmosphere are in a state of extreme dryness, but the law of decrease is thought not to be a regular progression. Professor Forbes considers it probable from many circumstances, "that the dryness is pretty constant for a certain height, and then rapidly diminishes." He adds, that "there is certainly a stratum of air at the height of from one mile to four miles, which is more frequently saturated with vapour than any other, and which constitutes the region of the clouds*."

(217.) This circumstance of the quantity of vapour being very different at different heights in the atmosphere, serves to explain many anomalies in the weather, while it shows the necessity of attending to the barometer as well as the hygrometer, in all our determinations respecting it. Thus, it often happens that the sky is clouded, and even rain falls, notwithstanding a low dew-point where the observer is situate. In summer, especially, thunder-storms frequently come up, while the air below, heated by the soil with which it is in contact, is comparatively dry. I have even known an instance, in which rain came on at 9^h 30^m A.M., and continued hard and steady, with scarce any intermission, the whole day, notwithstanding the dew-point, when it commenced raining, was 13° below the temperature of the air: the two previous days had been very fine, but the barometer falling; and it was a sudden change of the wind, with a consequent mixing of currents, which so completely, and in a very short time, altered the vaporous condition of the atmosphere above, before the effect could be communicated to the air below.

(218.) Sometimes, on the other hand, the early part of the day will be much clouded, with a very high dew-point, but by reason of a drier air prevailing at the elevation of a few hundred feet, the excess of vapour,

* *Report of Brit. Assoc.* 1832, p. 244.—Mr. Welsh, in each of the four balloon ascents, alluded to in a former part of this work (p. 69, note), found the humidity of the air "to increase till reaching the first stratum of clouds, afterwards to vary irregularly."—See his paper in *Phil. Trans.* 1853, p. 311.

as the temperature of the day advances, is taken up by the ascending current and dissipated, and instead of rain ensuing, the sun breaks out and the afternoon is very fine.—" In winter," also, occasionally, " the air is very moist for a considerable time, without rain; chiefly during the prevalence of foggy days and frosty nights, with a high barometer*." And it is this last instrument, which, in most instances, will speak to the true character of the weather, when the hygrometer gives a doubtful indication.

DEW.

(219.) The researches of Wells on dew have left little to be determined by future observers†. His theory, indeed, has been disputed by some few meteorologists, but it has more generally been accepted, as, on the whole, better explaining all the facts connected with this phenomenon than any other that has been offered. Wells has clearly shown that dew is owing to the gradual condensation of moisture on the surfaces of bodies on or near the ground, cooled down by radiation at night below the temperature of the air. If the temperature descend to the freezing-point, the dew appears as *hoar - frost* ‡. The conditions necessary for the

* Howard, *Climate of London*, vol. i: p. 91.

† *Essay on Dew*, 1818.

‡ Dr. Davy has observed that the temperature of dew sometimes falls to or below the freezing-point, without being frozen. He witnessed this on three occasions, when the thermometer on the grass was 29°, 32°, 30°, respectively. " In the first instance, it was observed at 12 at night, when the air was very calm as well as clear; observed an hour later, the thermometer on the grass had fallen to 27°, and the dew was frozen." It was " conjectured that the dew in this and the like instances was preci-

formation of dew in any quantity, are a humid and still atmosphere, and a clear sky favourable for radiation : the situation, also, must be tolerably open to the sky, and not sheltered by trees or other surrounding objects. If the air is not to a certain degree moist, there will be no condensation, and consequently no dew. If it is not still, or at least if there is much wind, the dew, as fast as formed, will be taken up again by currents passing over the spot, in like manner as agitation of the air promotes evaporation under any other circumstances. If the sky is not clear, the clouds in proportion to their density, as an interposing barrier, check the radiation of heat upwards from terrestrial bodies, while they themselves radiate a certain amount of heat downwards to the earth : thus bodies on the ground are prevented from being cooled below the temperature of the surrounding air.

(220.) Other conditions remaining the same, dew will be most plentiful on those evenings when the air is most humid. Heavy dews, therefore, in general forebode rain; since they indicate an amount of vapour in the air approaching saturation.

(221.) Occasionally, circumstances may favour the continued deposition of dew throughout the night; but since radiation exerts its greatest power about sunset and sunrise (85), it is at such times that the dew will

pitated at a temperature a little above freezing-point, and that it remained unfrozen, owing to the great stillness of the air, after its temperature had been reduced below the point of congelation, much after the manner of water confined in small tubes at rest, on which, without freezing, the temperature has been brought as low as 20° and 17°."—*Edinb. New Phil. Journ.* vol. xxxix. p. 15.

collect in the largest quantities*. When the sky is changeable, it may form and be dissipated, and re-form several times in succession.

(222.) Dew first forms at the roots of grass, or on those portions of the stem and blade which are nearest the ground. If, in the afternoon, when the sun is getting low, and the temperature beginning to decline (supposing the weather favourable for the formation of dew), we examine with a lens the fine grass on a mown lawn in shaded places, it is easy to trace this phenomenon from its commencement. We shall first see extremely minute globules of water, not visible to the naked eye, and even when magnified scarcely amounting to more than a dull film, collecting at the bottom of the herbage, the upper portions still remaining bright and dry. As the process goes on, the dull film spreads upwards, while the minute globules at bottom become larger, and the size of the drops keeps continually increasing, as the evening advances, and the temperature gets lower, until the whole stem and leaves are bedewed alike.

(223.) This circumstance of the dew appearing first towards the roots of grass, is owing, not to the cold induced by radiation commencing there, but to the greater humidity of the stratum of air in immediate contact with the soil. The cooling process commences at the extremities of the blades, which, from being more exposed, are the first to radiate out the heat they have received during the day. From them the cold is speedily communicated to that portion of the atmo-

* According to Harvey, dew sometimes continues to be deposited till near *two hours after sunrise.*—See *Quart. Journ. of Sci.* vol. xvi. p. 41.

sphere by which they are surrounded : more gradually
it is extended downwards to the earth. If a thermometer be placed at the roots of grass, the temperature
of the ground will often be found higher than that of
the air a few inches above it, for some time after the
dew has begun to form; and so long, at least, as this
is the case, the earth continues to emit a vaporous exhalation, as during the day, though in less quantity.
The lowermost stratum of air receiving more of this
vapour than the strata above it, is sooner brought to a
state of saturation, and ready to have some part of its
moisture condensed upon the grass at a higher temperature than the latter. It is probable that a considerable portion of the dew first formed, arises from this
condensation of the earth's vapour; while later in the
evening, as the cold increases, the exhalation from the
earth being much diminished, if not wholly checked,
the dew is then chiefly caused by a condensation of
the watery vapour contained in the atmosphere itself.
Wells thinks that on cloudy nights, " all the dew that
appears upon grass may sometimes be attributed to a
condensation of the earth's vapour ;" but that on nights
favourable to the production of dew, this is not generally the case*.

(224.) It must be remembered that cold is the
cause, not the *effect* of dew. " Cold dewy nights,"—
such as often occur in spring and autumn, when the
temperature at night falls considerably below what it
has been in the day,—is an expression often used by
agriculturists and gardeners; but the nights are not
cold because dewy, as these persons suppose, but the
reverse, dewy because cold, the atmosphere at the

* *Essay on Dew,* p. 243.

same time being humid, and favouring a deposition of moisture.

(225.) Dew is not deposited alike, or in equal quantities, on all bodies: it depends upon their radiating properties. Those which are the worst conductors of heat, are the best radiators; and it is on these last, such as grass, cotton, wool, and other filamentous substances, that dew is deposited in the greatest abundance; while on the good conducting surfaces of metals, as also upon the naked soil, it is rarely if ever produced.

(226.) " In remarking," however, " that dew is never formed upon metals, it is necessary to distinguish a secondary effect, which often causes a deposition of moisture upon every kind of surface indiscriminately. The cold which is produced upon the surface of the radiating body, is communicated by slow degrees to the surrounding atmosphere; and if the effect be great and of sufficient continuance, moisture is not only deposited upon the solid body, but is precipitated in the air itself, from which it slowly subsides, and settles upon everything within its range*."

(227.) Sometimes we hear persons speak of dew as "rising," while others speak of it as "falling;" but neither of these expressions is quite correct. If we restrict the word *dew* to the actual globules of water as they appear on the leaves of plants and other bodies, these globules do not exist until they are deposited by condensation of the aqueous vapour in that stratum of air in immediate contact with the surfaces of those bodies: thus they neither rise nor fall. But if we speak of the dew collectively, and in reference to the

* Daniell, *Meteorology*, vol. ii. p. 520.

successive appearance of these globules at a continually increasing distance from the earth, as above described, the expression of "the dew rising" would not be so very inappropriate. We then simply mean that the dew is extending more and more upwards, though no individual globule rise higher than the point at which it was first formed.

(228.) Again, by "falling dew" a phenomenon is sometimes meant to which the name of *dew* is scarcely applicable, though it might perhaps be called a falling *mist*. I allude to that cold damp state of the atmosphere, which at certain seasons of the year is due to the descending *cumulus*, or day-cloud, towards evening. When the vapours, to which these clouds owe their origin, cease to be supplied from the earth, by reason of the declining heat, evaporation begins to take place at their lower surface, and they are gradually absorbed. But this evaporation causes cold, in consequence of which that portion of the air in which it is going on becomes heavier, and sinks to a lower level. "As evaporation of the cloud proceeds, greater cold is produced, and by the time that the whole of the cloud is evaporated, the mass of air is so much cooled, as frequently to become heavy enough to sink to the surface of the earth, where it constitutes the cold air that is often felt in the evenings succeeding warm days in the summer and autumn. When the day-cloud is very large, the atmospheric mass is sometimes sufficiently cooled to cause it to descend to the surface of the earth, before the globules of the water constituting the cloud are all evaporated. These globules are then found floating in the lower air, and any object passing through them is soon wetted by them, as if by rain, though they do not,

like drops of rain, fall freely to the ground*." It is
this vaporous state of the atmosphere, which, in the
above kind of weather, dims the lustre of the declining
sun, causing it to appear of a dark purplish red, the
tint deepening as the sun approaches the horizon.

(229.) Several phenomena analogous to dew, and
capable of a similar explanation, may be observed in
houses, where they are familiar to most persons, though
not always traced to their right cause.   Such is the
dew on windows, when the temperatures in- and out-of-
doors are very different, and which is usually spoken of
as the windows being " steamed." The " steam," or
dew, is sometimes on the inside of the panes, and
sometimes on the outside. The former is the case on
cold autumnal mornings, and all other occasions, when
the temperature of the outer air is much below what it
is in the house ; the glass is then cooled down partly
by radiation, and partly by being in contact with the
external air, and the moisture contained in the warmer
air of the rooms is condensed upon the panes. If the
temperature out-of-doors fall below the freezing-point,

* See Hopkins on " The Formation of Dew," in his work *On
the Atmospheric Changes which produce Rain and Wind, and the
Fluctuations of the Barometer,* 2nd Edit. 1854, p. 199.

The above kind of "falling dew," if we choose to give it that
name, is alluded to by White, and not incorrectly described
by him as follows :—" After a bright night and vast dew, the sky
usually becomes cloudy by eleven or twelve o'clock in the forenoon, and clear again towards the decline of the day. The reason
seems to be, that the dew, drawn up by evaporation, occasions
the clouds ; which, towards evening, being no longer rendered
buoyant by the warmth of the sun, melt away, and *fall down
again in dews.* If clouds are watched in a still, warm evening,
they will be seen to melt away, and disappear."—*Naturalists'
Calendar* (by Aikin), p. 145.

the moisture is congealed, and we say " the windows are frozen."

(230.) Sometimes the moisture on the inside of windows is caused by a sudden storm of hail, or cold rain from the North, in summer, rapidly lowering the temperature of the external air; or it may be produced by an apartment being over-heated, as when a number of persons are collected within it, their united breaths at the same time making the air of the apartment very humid.

(231.) When the moisture is on the outside of the window, it is due to the outer air being warmer than the inner,—the converse of the above. This often happens in winter, on the occasion of a sudden thaw after a protracted frost, whereby the rooms of a house have been considerably cooled down. The thaw being attended by humid Westerly or South-westerly winds, so soon as these last come into contact with the cold panes of glass, the vapour which they contain is condensed upon them. On the same occasions, we sometimes see the condensed vapour standing in drops upon the cold walls of a house, or even running down in streams, if the outer door has been left open for any time, allowing the warm air to enter. This occurrence is often described as "the damp coming out of the walls;"—but it is quite a mistake to suppose that the moisture existed in the walls previously; it exists in the air, and is simply condensed upon the surface of the walls, just as the dew is condensed upon the grass and other bodies out-of-doors, cooled down below the temperature of the surrounding air, or as we cause a dew to appear on a mirror when we breathe on it.

(232.) There is another phenomenon which may be

mentioned here, though not so immediately connected with the subject. We must often have observed the slipperiness of roads, on the first approach of a thaw, beyond what occurs during the uninterrupted continuance of the frost. This is especially the case when there has been previously a moderate fall of snow, which after much traffic upon the roads, becomes beaten and dusty, and is travelled over by men and horses with little inconvenience; but so soon as the thaw comes, the surface of the snow, instead of liquefying and giving way in the first instance, becomes smooth and glassy, and so slippery, that it is difficult to walk, or in some places even to stand upon it with safety. In fact the surface is like that of a regular glacier. This circumstance is sometimes due to the thaw being accompanied by light rain, which is frozen into ice on coming into contact with the snow lying upon the road. But the same thing occasionally takes place without rain; and it is then caused by the vapours, with which the comparatively warm South-westerly winds are charged, being condensed upon the surface of the cold road, and the latter thereby covered with a coat of ice.

CLOUDS.

(233.) When the quantity of vapour in the air exceeds what can be maintained in a transparent state at the existing temperature, a portion of it is precipitated. The precipitation may take the form of dew, cloud, mist, rain, hail or snow, according to circumstances. Dew, which has been already spoken of, is a precipitation of vapour from that stratum of air in immediate contact with the earth, and with the surfaces of those bodies upon which the dew is deposited. The cloud is a visible aggregate of very minute aqueous particles, precipitated in the air itself, and suspended at a certain elevation above the earth. It assumes different forms dependent upon the particular conditions under which it exists; these conditions being probably electrical*. If the precipitated vapour, instead of being collected into masses, is generally diffused through the air, we call it a mist or fog; which may vary in density from a slight haziness, to a fog so thick as to allow of only very near objects being seen through it. In the case of both cloud and mist, "the humidity that is in the air is slowly parted with; when it is more copiously detached," and in drops sufficiently large to overcome by their weight the resistance of the air, and to gravitate to the earth, "it forms rain or hail†."

(234.) The aqueous particles, of which clouds and mists consist, have been thought by Halley and others to be hollow vesicles, "the water only serving as an envelope." Kämtz, however, regards these vesicles as "probably mixed with a great quantity of drops of water‡." The size of the vesicles of fog is very vari-

* "Of the cause of the formation of clouds, there seems no theory more probable than that which refers them to the union of atmospheric strata of different temperatures, unequally charged with moisture. At the time, however, in which this union takes place, electricity is evolved, disposing the watery particles to assume a vesicular form. The vesicular atoms, *charged with the same kind of electricity*, cause a general repulsion among all the vesicles, imparting, probably, to clouds their peculiar forms, and preventing them from descending in the shape of rain."—*Encyclop. Metrop.*, art. "Meteorology."

† Leslie.

‡ See a paper, in connexion with this subject, by Dr. Waller, who thinks that the so-called vesicles of fog and cloud (or

able, and dependent upon season. " In winter, when the air is very moist, their diameter is twice as great as in summer, when the air is dry." Their size also varies in the same month, according to the degree of the humidity of the atmosphere.

(235.) Each kind of cloud seems to have its fixed limits of elevation, within which it ordinarily keeps. The highest clouds are the fine white streaks, which appear on the face of the sky in serene weather, and which Dalton found, by several careful observations, to be from 3 to 5 miles high. As the temperature of the air at that great elevation must necessarily be many degrees below the freezing-point, it is probable that these higher clouds are composed of frozen particles: Kämtz thinks that they are composed of flakes of snow*. The other forms of cloud take a much lower range; most of them, in the opinion of Gay-Lussac, floating at an altitude averaging from 1500 to 2000 yards. Thunvesicular vapour) consist entirely " of minute globules and spherules of water."—*Phil. Trans.* 1847, p. 23.

* Kämtz infers this from the circumstance of halos and parhelia being formed among these lofty clouds, which phenomena "being due to the refraction of light in frozen particles," he concludes that the clouds in question " are themselves composed of flakes of snow floating at a great height in the atmosphere." He adds, " Observations, continued for ten years, have convinced me of the truth of this assertion ; and I know of no observation tending to prove that these clouds are composed of vesicles of water. We may feel astonished no doubt, that in summer, when the temperature frequently attains 25° (77° Fahr.), the clouds, which float above our heads, are composed of ice; but the doubt will disappear when we reflect on the decrease of temperature with height. During one of those hot days, when rain falls on the plains, this rain is snow on the summits of the Alps."—*Meteorology* (by Walker), p. 119.

der-clouds occasionally are seen very low: when very heavy and black, they sometimes appear almost to touch the earth at their lower portion.

(236.) It is hardly possible in all cases to refer the clouds to any definite forms; nevertheless the system and nomenclature proposed by Howard, and adopted by most meteorologists, serve well to characterize certain principal forms, of which all others appear to be modifications. And though these forms may be dependent upon changes going on in the cloud itself, which we do not thoroughly understand, and the system so far fall short of conveying to us all the knowledge we could desire, yet the several forms themselves at least assist in indicating weather-changes, when compared with anterior states of the sky and atmosphere.

(237.) Howard enumerates seven principal kinds of clouds, to which he gives the following names:— 1. *Cirrus*; 2. *Cumulus*; 3. *Stratus*; 4. *Cirrocumulus*; 5. *Cirrostratus*; 6. *Cumulostratus*; 7. *Nimbus*. As these clouds, with their several varieties, have been so often described in meteorological works, I shall not do more than make a few remarks illustrative of their respective appearances.

(238.) The *Cirrus* is a name given to those light streaky clouds of a thread-like texture, already alluded to (235), which exist at a great elevation, and are often the first clouds that make their appearance at the approach of change, after a long run of fine weather. They are very changeable, and assume various forms, the fibres of which they are composed being sometimes straight, sometimes flexuous or curling, sometimes divergent from a nucleus in all directions.

(239.) The *Cumulus* is so called from the large dense

masses into which it collects, the heap increasing upwards from a horizontal base.   It is essentially a day-
cloud, forming gradually in the morning a few hours
after sunrise, attaining its largest size in the hottest
part of the afternoon, declining after that period as
gradually as it formed, and finally disappearing about
sunset.   This cloud, in its most perfect form, is the
usual attendant on fair summer weather, and is seldom
seen after October.   It seems dependent upon temperature; and is evidently due to the vapours drawn upwards from the earth by the action of the sun, in increasing quantity as the temperature advances.   When
the cumulus becomes very large and of irregular shape,
or is about to pass into the form of cumulostratus,
mentioned below, it is indicative of showers.

(240.) The *Stratus* is that sheet-like mist, which is
seen reposing on or near the ground, in valleys and
damp meadows, towards evening, and sometimes throughout the night.   In this respect, it is the direct opposite
of the cumulus, or day-cloud, which last frequently
takes its origin from the stratus gradually ascending
into the atmosphere after sunrise, at the same time
altering its form.   More will be said of the stratus
further on, under the head of Mists.

(241.) The *Cirrocumulus*, as its name implies, is
somewhat intermediate in form to the cirrus and cumulus, deriving its origin from the former, of which it
may be considered a modification.   It comprises those
small patches of fleecy clouds, which are often seen in
great numbers in summer at a considerable elevation,
though not so high as the cirrus itself, and which are
generally the precursors of heat as well as of rain : they
are, however, sometimes seen in winter, and in the

intervals of showers. The dappled form of sky called *mackerel-backed* belongs to this formation.

(242.) The *Cirrostratus* is one of the most variable of clouds, assuming almost every possible form, and only constant in respect of its greater density and less elevation, generally, than either the cirrus or the cirro-cumulus. It constitutes the dark cloudy sky which precedes a change to wet, and, when seen in any great abundance, is always indicative of unsettled weather. Or it is more thinly diffused over the heavens, like a fine sheet, through which the sun is scarcely able to make its way. It is also seen in the intervals of storms, accompanied by cirrocumulus and other modifications. Sometimes it takes the form of bars and streaks, having the appearance of a shoal of fish. It is generally this cloud through which the moon is seen, when the latter is surrounded with a burr or halo, and it is on this account that the halo is rightly considered as a prognostic of wet.

(243.) The *Cumulostratus* is a modification of the cumulus, and is due to the rapid enlargement of this latter at top, where it passes into the form of cirrostratus. Mountain-masses of this cloud may be seen sometimes, of such size, and so closely compacted, as almost to cover the sky; attesting to the rapid accumulation of vapour set free in the air, and to the probable occurrence of showers, or hard storms. Belville observes that "the cumulostratus forms the basis of great thunder-storms, its electrical character attracting clouds and scud from all quarters of the heavens, which, uniting confusedly, constitute that indescribable black mass always antecedent to storms of thunder and lightning*."

* *Manual of the Barometer*, p. 30.

K

Such storms come on most frequently about 2 or 3 o'clock in the afternoon, that being the hottest time of the day, when the quantity of the ascending vapour is at its maximum. The cumulostratus is most often observed in spring and summer.

(244.) *Nimbus.*—The cloud so called is one in the act of being precipitated in the form of rain, and is the only one in which the rainbow is seen. It may be considered as a distinct modification, so far as it arises, not merely from the union of two or more of the preceding modifications, but from the clouds so united going through certain changes, due probably to a disturbance of their electrical condition, and causing the peculiar appearances that may be observed on the first coming on of rain, as seen in profile in a shower in the distance. Until these changes commence, no rain takes place, notwithstanding the sky may be completely covered with clouds of different kinds, and of the most threatening aspect. The following extract from my Meteorological Journal affords an instance of the appearances presented by the clouds before rain, which appearances, however, cannot be distinctly noticed, when the sky is completely overcast for some time previous to the rain actually commencing.

Nov. 21.—Fine at sunrise, but, soon after, the sky began to exhibit very curious features of cirrocumulus. In some places this formation displayed itself in moderate-sized, and very closely compacted, yet quite distinct masses of cloud. In other parts of the heavens these masses gradually decreased in size, as well as density, till they resembled mere specks, occasionally somewhat tufted, and gave the sky a peppered appearance. The whole of these were at a great altitude, and almost stationary; but *beneath*, there occasionally floated along, in a brisk current of air, thin sheets of cirrostratus, varying, however, in density.

Towards 9 A.M., these last had much increased; and, shortly after, they inosculated with the upper stratum of cirrocumulus, and nimbus ensued, the rain continuing, at times very heavy, all day. Barometer (when the rain commenced) 29·70. Temperature of the air 54°. Wind S.W.

(245.) The character of the clouds, and the exact circumstances under which they pass into the form of nimbus, may not always be the same as in the above instance. There are many other modifications of this phenomenon. But in all cases of settled rain,—no less than in the case of passing showers, in which the process (when the shower is forming in the distance) may be more easily observed,—there appear to exist these two strata of clouds, one above the other, with a considerable interval between; which strata, by their inosculation, constitute the nimbus; the upper one consisting either of cirri, or cirrocumuli,—the lower one either of cumulus passing into cumulostratus, or cirrostratus. Further, these two strata continue to maintain their relative position after the several forms themselves have been blended and lost. So long as the rain continues, the upper one is mostly concealed; but when the lower stratum breaks, previously to the original forms of cloud reappearing, the upper one comes into view. When the upper stratum breaks as well as the lower, letting in the sun and blue sky, we judge that the weather is about to clear, and that the rain will soon cease*.

* The above two superimposed strata of clouds during rain have been described by Howard and others. Mr. Monck Mason, in particular, has recorded several details ascertained by balloon ascents, relating to the condition of the sky above the clouds, from which rain is falling. He says, "whenever from a

(246.) If the rain from a nimbus fall into a stratum of air drier than that in which the cloud is formed, it may be taken up again in the act of descent, and not reach the earth at all. This may be often observed in the case of light showers in the distance at a certain elevation, while the sky nearer the horizon is quite clear.

(247.) Showers of rain, when seen some way off in profile, have the appearance of being stationary, from our not being able to distinguish the individual drops, the constant succession of which in a downward direction looks like so many parallel dark lines in a fixed position : it is not until the rain reaches the spot where the observer is situate, that this appearance goes off. Somewhat similar to this, is the apparent stationary character of clouds on the summits of mountains, notwithstanding a strong wind may be blowing at the time, a circumstance which often surprises travellers who are not familiar with the occurrence. In this case, however, the deception arises from the precipitated vapour, which is really borne along in a horizontal

sky completely overcast with clouds rain is falling, a similar range of clouds invariably exists in a certain elevation above, whereby the rays of the sun are intercepted from the layer below; and, on the contrary, whenever, with the same apparent condition of the sky below, rain is altogether or generally absent, a clear expanse of firmament, with a sun unobstructed by clouds, is the prevailing character of the space immediately above : thus leaving it a determinate fact, that when rain is pouring from clouds overspreading the earth, the rays of the sun are not operating upon the clouds in question; while, on the other hand, rain does not fall from such clouds when the rays of the sun are unobstructedly falling upon the upper surface."—*Rep. Brit. Assoc.* 1841, p. 56.

direction, continually forming on one side of the mountain as fast as it is dissolved on the other. The vapour comes from the warmer stratum below, which is forced up along the flanks of the mountain, by the resistance this latter offers to the horizontal current, and is condensed at the top; but no sooner has it been carried by the wind a short distance from the summit on the other side, than it falls again into the warmer stratum, and is dissipated.

(248.) In addition to the above seven principal forms of clouds, *scud* may be mentioned,—a name given to those dark shapeless masses of cloud, which are hurried rapidly along before the wind, at a less elevation than any other clouds except the stratus, more especially after rain, and during unsettled weather in winter. In many cases they appear to be as it were the broken-up fragments of some storm, which has nearly spent itself, and previous to its entire dispersion*.

MISTS.

(249.) The terms *mist* and *fog* are often used indiscriminately to signify the same phenomenon. It may be convenient, however, to restrict the former name to those low creeping mists, which rise but a few feet above the earth, and to apply the latter to mists reaching to a greater elevation, and generally diffused throughout the atmosphere. The former, however, which mostly occur at night, and are the same as the *stratus* cloud

---

* Those who desire further information about clouds, are referred to Howard's *Climate of London*, 2nd Edit. vol. i. Introd. pp. xxxix–lxxii.; and to Forster's *Researches about Atmospheric Phenomena*, 2nd Edit. Chapters I. and II. In this last work will be found representations of the different forms they assume.

before spoken of (240), often pass into the latter towards
morning, and occasionally prevail throughout the succeeding day.

(250.) Mists, taking the term in the above restricted
sense, assume a different character, according to the
circumstances under which they are formed. In flat
countries, their appearance is like that of an expanded
sheet of water, spread equably over the meadows, where
these are low and marshy, the trees and other high objects standing out of the mist, while everything near
the ground is concealed by the dense veil drawn over
it. When the country is undulated or hilly, it is in
the valleys alone that the mist forms at sunset; and
these are seen filled with mist the following morning,
which keeps rising higher as the day advances, until, in
fine weather, it is dispersed, and taken up by the warmth
of the sun. Occasionally, again, the mist, instead of
covering the meadows, follows the course of some river
or stream, over which it hangs to the height of a few
feet, marking out its course in the most distinct manner
to the eye of the observer, when otherwise, perhaps, its
windings would not be within sight.

(251.) These three different cases of mist require to
be explained in different ways; at least they cannot be
traced *directly* to the same cause. The mist over the
low meadows is primarily due to radiation, or the same
cause which produces dew. The grassy surface is first
cooled down as the heat of the day declines; this cold
is communicated to the stratum of air in immediate
contact with the earth; from thence it is propagated
to the next stratum above, and so on upwards, until
each stratum in its turn, beginning with the lowermost,
is cold enough to cause a precipitation of the vapour

which it contains: the mist then begins, and rises upwards in proportion as the cold increases.   Hence dew
and mists often appear together: but they are not
necessarily connected.   A dew, indeed, must always
precede the formation of mists; and will be most
copious, as Wells has observed, "on those clear and
calm nights, which are followed by misty or foggy
mornings."   But the converse does not hold true: a
mist does not always follow dew.   On many evenings
dew will be deposited, notwithstanding the air is transparent even to the earth's surface.   This arises from the
herbage and other bodies, upon which the dew is deposited, being cooled down by radiation to a greater degree, or more rapidly, than the air, generally, is cooled
down in the manner above described: the temperature
of this latter, therefore, does not fall to the point of
saturation, except where it comes into immediate contact with the surfaces of those bodies.   Whether the
mist follow or not, depends upon the quantity of vapour
which the air contains.   Hence a mist will often appear
in damp places, while in others, where dews are of
constant occurrence, a mist, *i. e. stratus*, may be a
rare thing.   It must be remembered, however, that
there are two distinct sources whence the lower strata
of the atmosphere derive their humidity as the day
declines.   The locality itself may be naturally damp,
from its proximity to a river or a marsh, or from its
low position combined with an excessive growth of trees
in the neighbourhood; or, though ordinarily dry at
other times, it may be damp at sunset (as most open
places are to a certain extent during clear weather),
from the descent of the cumulus, or day-cloud, before
spoken of when treating of falling dew (228).   Of

course, when these two causes combine, a mist will
much more frequently supervene upon dew, than when
there is only the latter cause to bring it about.

(252.) The kind of mist I am now referring to is, as
might be expected, common in Cambridgeshire, at least
in those parts of it bordering upon the fens.   At
Swaffham Bulbeck it is of very frequent occurrence;
and it has been a matter of much interest to me to
watch the first forming and gradual spreading of the
stratus, on the evenings of such days as were favourable
for its appearance, in the meadows adjoining the field
in which the vicarage house stands.   One year in particular, each time this mist occurred, I noted down all
the attendant circumstances of weather, and other conditions under which the mist first showed itself, as well
as of the changes this last occasionally underwent,
apparently from some slight alteration of those conditions.   It is from these notes that the following remarks are taken in illustration of this phenomenon.

(253.) Stratus generally occurs at or a little after
sunset: sometimes it does not come on till late at
night, disappearing a considerable time before sunrise;
at other times it only shows itself a little before sunrise, disappearing soon afterwards.   Though more frequent in the autumn, it is occasionally seen on the
evenings of the hottest days in summer.   I have known
a thick one at sunset, on the 14th of July, after a maximum temperature of 83°.   Its occurrence is either
during fine weather, being preceded as well as followed
by a fine day, or during changeable weather, when a
bright clear evening follows a wet morning.   In the
former case, the barometer is generally high, the thermometer, at the time of the stratus coming on, low,

compared with the maximum of the day, and the wind somewhere to the N. of East and West ; and the mist is mainly due to the difference between the temperatures of day and night. In the latter case, the barometer is often low, and the thermometer comparatively high, the wind in no particular direction, but the air very damp ; and it is the excess of moisture in the air, rather than any considerable fall of the temperature, which evidently causes the mist in this instance. Sometimes the barometer and thermometer are both high, but then the mist seldom comes on till near midnight. The conditions necessary for the forming of stratus are the same as those requisite for the forming of dew, viz. a humid atmosphere, a clear sky, and a still air ; the immediate cause being cold induced by terrestrial radiation.

(254.) Stratus is generally bounded upwards by a well-defined line, above which the atmosphere is clear. When it appears at sunset, if the conditions under which it forms continue unchanged, it goes on increasing and rising higher, as the night advances ; and if the air be very humid, or the season of the year late, it often leads the next morning to a regular fog uniformly diffused through the air. The way in which stratus gradually spreads itself sheet-like over a meadow, or at other times extends in lines and bands from one meadow to another, is very striking. If the meadow be equally low and marshy everywhere, the stratus will appear simultaneously over all parts of it alike, resembling at first a thin layer of cotton wool, or a light smoke rising from the ground and just level with the top of the grass, but quickly becoming denser and rising higher as the cold increases. But if one spot in the meadow be damper

K 5

than the rest, or one meadow more marshy than another
adjoining, the mist is sure to show itself first in such
situations, whence it travels to other spots or other fields,
having the appearance of being borne along by the
wind, the direction of which it often takes.   This,
however, is a mere deception : the mist once formed is
everywhere nearly stationary, and the contrary appearance arises simply from different portions of the air
being successively cooled down to the point of condensation, according to the quantity of vapour they contain;
the least humid, or those most distant from the spot
where the mist begins, requiring the cooling process,
through the influence of terrestrial radiation, to be continued the longest before that point is attained.   The
deception is, in fact, analogous to that of the apparent
rising of dew before alluded to (227), or to the rising of
the stratus itself from the earth when first forming; the
only difference is, that, in the two last instances, the
condensation takes place at successive intervals in an
upward direction,—in the instance above spoken of, in
a horizontal one.

(255.) It is also remarkable how, after stratus has
formed in the usual way, and risen gradually upwards
to the height it may be of eight or ten feet from the
ground, it will occasionally change its place very rapidly,
appearing suddenly in spots where it had not been
seen before, and disappearing in others, where but a few
minutes previously it was quite dense.   Sometimes it
will be suddenly dissipated near the ground, the upper
part of the sheet still remaining unaltered; or it will
appear in long streaks and bands, stretched horizontally
across the field, and elevated at different heights above
the ground; or it will go through other changes and

assume other forms.   This capriciousness in its apparent movements must be evidently due to slight fluctuations either in the temperature or the humidity of the air at the spots where it occurs.   If the temperature of a given stratum of air be suddenly raised, having previously been coincident with the dew-point, the vapour which had been precipitated, forming the mist, will be as suddenly taken up, and the air become clear.   If the temperature afterwards fall back again to what it was before, the mist will immediately reappear.   If the temperature of the air, having been coincident with the dew-point, vary but little, while the moisture in the air be first lessened, and then restored to its original quantity, the dew-point, in accordance with these changes, will first fall below, and then again rise up to that temperature, and cause first a disappearance, and then a reappearance of the mist.   And these changes may succeed each other several times over.   Are there, then, any causes at hand liable to bring about such fluctuations in the temperature and humidity of the air?   I conceive they are dependent upon the state of the sky, or due to very slight currents in the atmosphere.   There is also a third cause, to which some of the changes observable in this kind of mist may be attributed, and to which I shall refer after speaking of the two causes first mentioned.

(256.) It has been already stated, that one of the conditions necessary for producing stratus is a clear sky.   It never forms when the sky is clouded.   And if, after the stratus has appeared, clouds come over, they, by immediately putting a check to terrestrial radiation, while themselves radiate a certain amount of heat downwards to the earth, tend to raise the tempe-

rature of the lower strata of the air, and to disperse the
mist. The stratus may be either totally dissipated in
this way, or only partially broken up, according as the
sky becomes entirely overcast, or merely a few clouds
pass over it. In the former case, the mist will be first
dissipated nearest the ground (where the cold was first
induced by radiation), and gradually disappear from
below upwards, in the same manner as it came on at
the beginning. In the latter case, when the clouds are
light and scattered, the mist may be often seen playing
about from one spot to another, as the former pass
overhead, so immediate is their effect upon the temperature, by now concealing, and now letting in again to
view, the expanse of the sky.

(257.) But the stratus will sometimes vary in situation and density, notwithstanding the sky continue
uniformly clear the whole time. The changes which
it undergoes in this case must, therefore, be referred to
some other cause. As an instance, I may copy the
following entry in my Meteorological Journal, under
the date of August 5, 1831.

" Dense stratus this evening after sunset, varying in character
repeatedly. One meadow in particular, a quarter of an hour
after it had been completely overspread by the mist, was again
clear and free from it ; and after another short interval, the
stratus once more showed itself as thick as ever. Sky uniformly
clear, and the air calm, during these observations."

Though the air is described above as calm, I believe
such changes in the stratus, when there are no clouds
to influence it, to arise, nevertheless, from slight agita-
tions of the air, due to currents so gentle as to be im-
perceptible except by their effects in this way. The

atmosphere is seldom so still everywhere around as to be entirely free from such currents. The mere alterations of temperature in the lower strata of the air resulting from terrestrial radiation, would of themselves lead to slight local displacements of the air itself. If one piece of ground be more open and exposed than another piece in the immediate neighbourhood, radiation would first begin in the former, and the moment the temperature of the superincumbent air were lowered from this cause, its density would be increased, and the equilibrium between it and the other adjoining portions of the atmosphere disturbed. Or some trifling shift of the wind itself may take place soon after the mist has formed. The slight currents generated in either of these ways would cause a frequent mixing of airs of different temperatures, and, according to their relative degrees of humidity, might lead to the forming of a mist, or to the disappearing of one that already existed*. If a current just below the point of saturation set in towards a spot where a mist is already formed, but be confined to a stratum of air somewhat higher than the upper boundary line of the mist, it would, on mixing with the air of that spot, if itself were of a slightly higher temperature than the latter, give rise to another thin line of mist, having a clear space between it and the one first formed. No doubt, also, in some cases, a mist may be dispersed near the ground by the setting in of drier currents, causing evaporation at its lower surface, while it still keeps increasing above, from a continued precipitation of moisture from higher re-

* Daniell has observed that "the agitation of merely walking through the condensed vapour forming the mist, is frequently sufficient to disperse or melt it."—*Meteor. Essays*, vol. ii. p. 524.

gions of the atmosphere. It is obvious such changes might be indefinitely varied.

(258.) It may seem rather paradoxical to affirm what, nevertheless, is in a certain sense true, that the air is drier during a mist than immediately before the mist comes on or after it disappears. But we must, in this case, distinguish between the air itself and the condensed vapour forming the mist. Previous to condensation, the vapour itself is in the aëriform state, and intimately, though still only mechanically, mixed with the air. After precipitation, the air is freed, as it were, of so much vapour as has assumed the aqueous form, though more remains in it, which may still be condensed by a greater degree of cold. Accordingly, if we take the temperature of the vapour in the air by a Daniell's hygrometer, before, during, and after a mist, we shall find that it is lower during the prevalence of the mist, compared with the temperature of the air, than it is at either of the other times; or in other words, the dew-point is lower, which is equivalent to the air being drier. Such at least appears to be the result of observations made by myself on several different occasions during the occurrence of stratus, some of which, in confirmation of this point, it may be desirable to record, just as they were noted down at the time.

(Obs. I.) July 14.—Thick stratus over some of the low meadows at 8^h 30^m P.M. Weather for some days before changeable. Close and sultry in A.M., with heavy rocking clouds and some rain. Wind S.W. Barometer at 10 A.M. 29·975. Temperature of air 65°. Dew-point 63°. In P.M. wind shifted to N^{ly}, and it became very fine and clear for some hours previous to stratus coming on. Barometer rose rapidly after the change of the wind, standing at 10 P.M. at 30·220.

At 9 P.M., when the stratus was in the adjoining meadow,

but not yet arrived at the field in which the observation was made, temperature of air was 55°, and that of the vapour the same, or coincident with it: radiating thermometer on the ground 52°. About a quarter of an hour after, when the mist had become general, the temperature of the air *in the stratus* (the same spot where the first observation was made) had fallen to 53°, and that of vapour to 51°; no longer coincident, but showing a difference of 2°: radiating thermometer 50°.

(Obs. II.) Oct. 19.—Thick stratus over the adjoining meadow at 6 P.M. Weather for some days previous settled and moderately fine, with high barometer, and wind S.W. A little rain, notwithstanding, fell yesterday evening, after which wind shifted to N. This morning rather overcast and cool. Barometer at 10 A.M. 30·257°; temperature of air 53°; dew-point 49° Sky cleared before sunset.

Exactly at 6 P.M., when, as before, the stratus was near at hand, but not yet at the place of observation, temperature of air was 40°, while the thermometer in the blackened bulb of the hygrometer stood at 39°, the black bulb itself being covered with a copious dew. The exact point at which the vapour had been first condensed upon the glass was not noticed, as the hygrometer had been left out to acquire the temperature of the air, and unfortunately not sufficiently watched. The bulb was then wiped dry, and its temperature raised by the warmth of the hand, after which it was again left to itself. At 6ʰ 15ᵐ, the stratus was everywhere diffused about the place of observation, and temperature of the air had fallen to 39°: the thermometer in the black bulb was at the same temperature, but the bulb itself was dry, and had evidently remained so since the wiping. A little æther was then employed to depress the temperature, and a ring of dew appeared at 37°, but the temperature fell to 33° before there was as copious a deposit of dew as there had been before at the temperature of 39°. Radiating thermometer on ground 36°.

At 6ʰ 30ᵐ, the stratus was entirely gone off again from the place of observation, though still hanging over the adjoining meadows. At this time the temperature of the air was 38°, and the black bulb of the hygrometer covered with dew : wiped the bulb, and raised its temperature, after which suffered the latter gradually to fall again, and a sensible dew reappeared at 39°·5. At 38° (the same temperature as the air) the deposition was copious. At 7ʰ, on again going to the hygrometer, found the temperature of the air 36°, and the thermometer in the black bulb 37°, with a copious deposition of dew : wiped the bulb as before, raised its temperature, and then suffered it to fall again. The dew reappeared at 37° exactly.

(Obs. III.) Oct. 25.—Dense stratus rising in the adjoining meadows immediately after sunset. At a quarter before 5, when the air was clear at the place of observation, but the mist very thick in the next field, temperature of air was 45°, and that of the vapour the same, or exactly coincident. The stratus afterwards varied very much in density at different times ; one minute thick over a given spot, at another wholly dissipated ;—never appeared at the place of observation at all. At a quarter after 6, the temperature of both air and vapour was 41° ; still coincident, but both fallen 4° : stratus still thick in some places, but none where the observations were made.

(259.) So far as the above observations go, which were conducted with great care to ensure correctness, they offer the following results. From the first and second cases, it would seem that just before the appearance of the mist, the dew-point is coincident with, *if not higher than*, the temperature of the air*, and from

---

* Since the above observations were made, I find that Daniell has mentioned the circumstance of the dew-point being, at such times, occasionally higher than the temperature of the air. He says, " In the most cloudless nights of the whole year, when the stars are bright, and the disc of the moon perfectly sharp and well defined, by bringing the hygrometer out of a warm room, it will be found that the point of deposition is often three or four degrees above the existing temperature of the air : proving thereby that particles of water, though invisible, are floating in the atmosphere."—*Lond. Journ. of Sci.*, 1822, p. 100. Also in his *Meteorological Essays*, he observes that "if the hygro-

the second it would seem that this is also the case after the mist has gone off; but that in the mist itself, or at least for a short time after its first appearance, the dew-point is lower than the temperature of the air by (in each of these instances) two degrees. In the third case, in which the mist never arrived at the place of observation, though thick everywhere about, the temperatures of the air and vapour remained coincident for at least an hour and a half, though during that time both fell 4°.

(260.) I would now suggest whether the circumstance of the dew-point rising again immediately after the disappearance of the mist, is not to be taken in connexion with that disappearance, and both explained by the giving-out of the latent heat consequent upon the condensation of the vapour in the first instance. Such condensation would tend to raise the temperature of the air at that spot, and enable it to take up some portion of the mist again. That the mist is not always wholly dissipated immediately from this cause, may arise from an alteration of some other of the attending circumstances, such as the continued increase of the cold, from the effects of terrestrial radiation, which may exert a greater influence than the heat generated by the condensation of the aqueous vapour, or the setting-in of slight currents which may impart to the air at that place an increased humidity. But though the mist sometimes hangs for hours together over a given spot, altering very little in character, or, it may be, increasing in density from one quarter of an hour to

meter be brought from an atmosphere of a higher temperature into one of a lower degree, in which condensed aqueous particles are floating, the mist will begin to form at a temperature several degrees higher than that of the air."—P. 151.

another, nothing is more common than to see it at other times (especially soon after its first forming) repeatedly appearing and disappearing at the same place, and at the same height from the ground, without any alteration in the face of the sky; and whenever this is the case, it seems not improbable that the liberating of latent heat in the above way, may, in conjunction with the other causes before alluded to (256) (257), play its part in bringing about those frequent changes in the pheno- menon in question.

(261.) *Mist in Valleys.*—What has been said hitherto on the subject of stratus, is in reference to those mists which appear in the marshy parts of flat districts. The mist that is sometimes seen filling the valleys of an uneven country, leaving the tops and sides of the hills clear, is to be explained by the circumstance of the air in valleys being not only more humid, but colder than on the hills, on those nights which are favourable to radiation. This at least is the case when the hills are "insulated, and of inconsiderable lateral extent." Ra- diation to a certain extent takes place on the hills as well as in the valleys; but in consequence of the air on the hill-tops being more agitated than that lower down, no portion of it rests sufficiently long in one place to become much colder, by "contact with the grass on the hill, than the general mass of the atmosphere at the same height." On the sides of the hills, again, the air is no sooner "cooled by contact with them," than, "from its increased gravity, it slides down their decli- vity, making room for the application of new and warm parcels to the same surface. The motion too, thus excited in the air, near to the sides of the hills, must occasion a motion in that upon the summit, which may, in some measure, account for" the greater agitation of the atmosphere there above spoken of*. From these causes combined, the valleys get more and more filled with cold air as the night advances, causing at the same time an increased condensation of the moisture which the air contains, in the form of mist.

(262.) It is due to the above circumstance of the air in valleys being colder at night than on the hills, that the effects of autumnal frost are first visible in the former situations, potatoes, dahlias, &c. being often blackened or killed there on cold nights, while the same vegetables in gardens higher up are found untouched. This is a matter of surprise to many persons; and even gardeners, who are aware of the fact, seldom trace it to the right cause. As these frosts are frequently accompanied by mist, there is often the same mistake here, as in the case of dew before alluded to (224). Persons think that the mist is the cause of the cold, instead of the cold being the cause of the mist. Whereas the mist rather tends to check the cold than otherwise, by impeding radiation, and by the vapour in the air giving out latent heat, while in the act of being condensed.

(263.) *Mist over Water.*—When a mist appears over a river or stream, or above a wide sheet of water, it is due not simply to radiation, which, nevertheless, is the primary cause in this case also, but to the slow intermixing of two airs of unequal temperatures. These mists are most prevalent during the autumnal months, " at which season the radiation of heat from the ground

* Wells: to whom the reader is referred for further details in explanation of the circumstance of valleys being colder, on clear nights, than hills.—*Essay on Dew*, pp. 223-226.

and from the water is very different, owing to the difference of their temperatures being then greatest. In autumn the temperature of the water is both by day and night nearly uniform at 40°, its point of maximum density; while the temperature of the ground is during the day much higher than 40°, and during the night often much under that temperature. The water in most cases occupying the lowest situations,—whenever, from the inequalities of the surface of the ground, or from any other cause, the colder air produced by radiation over the ground, is made to mix itself with the warmer air over the water, the moisture in the warmer air is condensed, so as to become mist*"

(264.) This same explanation will serve to account for the mists that are so often observed over the sea in certain states of weather, in the colder months. Here also there is an intermixture of air of different temperatures. The air over the sea being warmer, as well as more humid, than the air over the land, if the latter be carried seawards by the wind blowing in that direction, it will, on mixing with the former, cause a precipitation of the moisture in this last, and bring on a mist.

(265.) In summer, after heavy rains, we may occasionally observe a mist, exactly similar to the stratus that forms at night, hanging over the ground *during the day*. This is due, however, to an opposite state of things to what occurs in nocturnal mists. In the case of the stratus over land at night, the surface of the ground is colder than the air. In that of the day

* Prout, *Bridgewater Treatise*, p. 282. See also Sir H. Davy's paper "On the Formation of Mists in particular situations."— *Phil. Trans.*, 1819, p. 123.

stratus, the soil having been previously much heated, causes, after the rain, an ascent of vapour, which immediately becomes condensed from the air being colder than the soil*. This kind of mist is analogous to, and well illustrated by, the steam which rises from the backs of horses, when heated by work in cold weather; or that which ascends from a vessel of warm water. In like manner, also, the moisture contained in the human breath is rendered visible, by condensation, in winter.

Fogs.

(266.) Fogs are due to different causes, according to the circumstances under which they occur. These causes are mainly the same as those which occasion stratus, or the nocturnal mist, only in the case of fog a larger body of the atmosphere is acted upon, so as to be obscured to a greater height above the ground.

(267.) Fogs are either the result of the stratus of the

---

* From some experimental inquiries relating to mists, conducted apparently with great care, by the late Mr. Harvey of Plymouth, it would seem that during mists at night also, the air is sometimes colder than the ground, and "that the quantity and density of the mist increase in proportion to the *excess* of the temperature of the ground *above* that of the air." Proximity to the sea had probably some influence in the cases of which he speaks, the mists arising from an intermixture of airs of different temperatures, much as in the case of mists over rivers. Be however this as it may, I feel confident that in all the instances of evening mist I have observed at Swaffham Bulbeck, the surface of the ground, at least at the *commencement* of the mist, is colder than the air immediately above it. Those who are interested in the subject, are referred to Mr. Harvey's papers in the *Lond. Journ. of Sci.*, vol. xv. p. 55, and in the *Edinb. Phil. Journ.*, vol. ix. p. 255.

preceding evening gradually diffusing itself through the
night in an upward direction, and are then due (like
the stratus itself) to the condensation of the vapour in
the air by cold acting locally on a damp and still atmosphere; or they are caused by the mixing of two currents of air of unequal temperatures; or they arise from
an excess of vapour flowing into the air at a given spot,
where it is condensed by the cold of that place without
any manifest change in the direction of the wind. Sometimes, in winter, a fog is occasioned by a double current,
the colder one from the N. or N.E. being uppermost,
and continually falling by its greater weight into the
warmer one from the W. or S.W. below, and condensing
the moisture contained in this latter. Such a fog may
prevail several days, in fact so long as the wind remain
unchanged.

(268.) The morning fog, which has its origin in the
stratus of the preceding evening, occurs at all times of
the year, but more especially in autumn, when, if the
barometer be moderately high at the time, it is usually
dissipated at or before noon, and is followed by one of
those beautiful days, which are so characteristic of the
above season. In winter, however, the temperature is
often insufficient entirely to disperse the fog, and it
continues through the day, increasing perhaps in density as the night approaches. Whenever this is the
case, it will generally be found that the barometer has
risen higher than it was in the morning, indicating that
the wind has worked more to the N.E., and that a
larger body of cold air is flowing in from that quarter
to condense more and more of the atmospheric vapour.
It is often observable, indeed, at such times, that the
higher the barometer gets, the thicker the fog; and it

seems not improbable that in some instances the increase
of pressure, in addition to condensation from a decrease
of temperature, may serve to bring about this effect*.
When the fog clears off, as the day advances, we see it
first rising from the ground, and getting continually
higher above our heads with the increase of the sun's
heat, then breaking up into large and ill-defined masses,
and ultimately, when it has risen to a certain elevation,
forming the regular *cumulus* cloud, if the temperature
be sufficient, or melting away altogether, the whole sky
becoming bright and cloudless.

(269.) Fogs more or less thick often come on very
suddenly in consequence of the wind shifting from a
colder to a warmer quarter, or *vice versâ*, and are evidently due in that case to the mixing of two currents
of unequal temperatures.   The change of wind may
originate either in the current next the earth, or in a
higher one; if the latter, the direction of the wind will
for a time remain unaltered, as shown by the vane,
though the lower current will probably in the end follow
the higher one, but what has taken place above will be
immediately manifested, not merely by the fog, but by
the falling or rising of the barometer, according as the
change of the upper current is towards the S., or away
from it.

(270.) On the occasion of one of these sudden fogs
at Swaffham Bulbeck, one summer's evening, after a
fine day and a high barometer, I found the temperature

* Daniell has observed that "the condensation of vapour
may be effected, not only by decrease of temperature, but by
increase of pressure."—*Meteorological Essays*, p. 53.   In the
second part of his work, he remarks that "fogs frequently accompany a very high degree of atmospheric pressure" (p. 505).

of the air, at 8^h 20^m P.M., 61°·5, and the dew-point 55°.
This was at the moment when the fog, which a short
time previous had been seen advancing in the distance
from the S.W., had reached the place of observation.
Later in the evening, at 10 P.M., the temperature of the
air had risen to 63°, and the wind, which had been N.W.
all the day, was found next morning S.W. The cause
of the fog was therefore obvious.

(271.) If fogs, occasioned by a shifting of the wind,
are attended by a low barometer (say not above $29\frac{1}{2}$
inches), they are likely to turn to rain very shortly.
When they occur suddenly during clear frost, they indicate a tendency to change, and milder weather. But
if there has been previously much snow and cloud as
well as frost, the fog is a sign of the wind working
more to Northerly, and is likely to be followed by
severer cold. Long frosts are often ushered in by fogs
in this way. That of the winter of 1829–30 was preceded by a dense fog of four days' duration, thicker at
times than any I ever witnessed out of London. This
fog came on very suddenly in the morning, after a clear
night and bright sunrise, attended by a change of the
wind from S.W. to N.E., though it was inferred, from
the state of the barometer, that a S.W. current still
prevailed in the upper part of the atmosphere. And it
was apparently due to this latter circumstance that the
fog was succeeded by a heavy fall of snow, after which
the sky cleared, the barometer rose, and the temperature fell considerably.

(272.) If, during the prevalence of a thick fog and a
Northerly or Easterly wind, the wind change to the S.
or S.W., the fog will speedily disappear below, though
the sky may remain clouded overhead.

(273.) Thick drenching fogs often prevail for days together (I have known them last nearly a whole week), in autumn more especially, after a long run of wet weather with Westerly and S.Westerly winds. Such fogs I have noticed to be attended by a change of the wind to N.E., acting as a condenser on a very damp atmosphere, but the barometer too high to allow of any further precipitation in the form of rain, beyond a light mizzle, into which the fog occasionally resolves itself. After the fog has entirely passed off, the weather usually becomes fine and settled.

(274.) Kämtz has observed, that though "the formation of fogs is often determined by the sudden lowering of temperature in the same place where the vapour of water is raised from the soil, this vapour may," notwithstanding, "be transported by the winds into colder countries, and be transformed into fog at a notable distance from its place of origin*." This seems to account for a thick fog, such as I have known to occur in winter for two whole days, after a period of unusually mild and dry weather, and previous to the setting-in of much wet, but unattended by any change of the wind (which had been S.W. or Westerly for a long time before), either at the time of the fog coming on, or for some days after. The temperature also remained nearly the same during the continuance of the fog, and for many days after the fog had gone off, rose higher than it had been before, notwithstanding the weather was still wet and unsettled. These circumstances would seem to indicate that the fog on this occasion was not caused, as in so many other instances, by the mixing of two currents, or by any general depression of the tem-

* *Meteorology* (by Walker), p. 115.

L

perature of the air, but by an excess of vapour brought
up by the long-prevailing S.Westerly winds, and beyond what could be maintained in a transparent state
at the temperature of the place. A precipitation, therefore, assuming first the form of fog, and then of rain,
occurred, after an interval. This is further shown to
have been the case by the state of the barometer, which
fell nearly six-tenths of an inch during the fog, and
was scarcely higher than 29 inches on the morning of
the day after it cleared off, and when the rain fell
heavily for a great many hours without ceasing.

(275.) Fogs, and thick ones too, sometimes, after a
brief continuance, disappear as suddenly as they came
on, in a manner not easy to be explained. As an example of this, I copy the following entry in my Meteorological Journal, under the date of November 16,
1823 :—

"Rather a singular phenomenon occurred today. After a remarkably fine morning for the time of year, with a bright sun
and a Westerly wind, about 1 P.M. a dark patch was discernible
in the horizon, approaching in the direction of the wind, and
wearing the appearance of a heavy storm in the distance. As
it came up, which it did with great rapidity, considering there
was little or no breeze at the time, it turned out to be only a
dense fog, which in the course of a few seconds enveloped the
atmosphere to such a degree as completely to obscure the sun,
that had till then been shining brightly, and to render indistinct
all objects that were not very near. After this sudden darkness
had prevailed for a few minutes, it began gradually to clear off,
and in less than a quarter of an hour the fog had entirely passed
over, and the sun shone again as brightly as before."

At the time of this fog coming on, the barometer
was very high (nearly 30½ inches), but fell slightly in
the evening, when the fog reappeared, followed by a

little rain. For several days after, the weather was cloudy, but the barometer still high, and no more rain ensued for a considerable period.

(276.) I have at different times, during the occurrence of those thick fogs that are so frequent on autumnal mornings, made observations on the temperatures of the air and vapour similar to those I have already recorded when speaking of stratus. The results seem to indicate, that at or about sunrise, and for a short time afterwards, the dew-point is coincident with the temperature of the air; but that, as the sun gets higher, a difference between the two appears, and slowly increases, until, about the time of the fog clearing off (if it clear off at all), and the sun coming out, it has amounted to from 3 to 5 degrees, according to circumstances. The fact therefore before mentioned, of the dew-point being lower in the evening stratus than it is just before the mist comes on, is probably confined to a very short period after its first appearance : if the mist is not dissipated by any of the causes which sometimes interfere to check its continuance, but go on increasing during the night, so as to form the fog on the following morning, it would seem from other observations of my own that the moisture in the air soon fills up again to the point of saturation, at which point it remains till the temperature of the air is once more raised by the returning influence of the sun above the horizon. The following extracts from my Meteorological Journal show some of the observations I have made with a view to these inquiries :—

(Obs. I.) Sept. 4.—Thick fog this morning at sunrise, lasting till 10 A.M. Two preceding days fine and settled, after a week's wet weather. At 7^h 30^m A.M., on exposing a Daniell's hygrometer in an open part of the garden, the temperature of the air was found to be 51°, but to oscillate very remarkably, being successively, in the course of a few seconds, 51°, 52°, 49°·5, 50°. By two observations on the temperature of the vapour, this last was found to be 49°. At the time of the first, that of the air was 50°; at the time of the second, 51°; in neither case, therefore, were the two coincident.

At 8^h 30^m, temperature of the air 54°; that of the vapour 52°

At 9^h 30^m, temperature of the air 58°; that of the vapour 53°. Fog now somewhat breaking up; disc of the sun just visible: a few minutes later, fog in large floating masses, blue sky visible, and sun out at intervals.

At 10^h, temperature of air 59°; that of vapour 55°. Barometer 30·271 inches. Wind W., a little variable. Fog gone, and cumuli forming over-head, with a bright sun; fine for the rest of the day.

(Obs. II.) Oct. 21.—Thick fog, preceded by dense stratus the evening before. At 7^h 15^m A.M., when the fog was everywhere very thick, the dew-point and temperature of the air, in an exposed part of the garden, were exactly coincident, both being 39°.

At 8^h, in another part of the garden nearer the house, the temperature of the air was 42°; dew-point 41°: fog as thick as before.

(Obs. III.) Oct. 26.—Thick fog at sunrise, preceded by dense stratus the evening before. At 7^h A.M., when the fog was very thick, and a hedge at the bottom of the garden barely visible, the dew-point and temperature of the air were exactly coincident, both being 42°·5.

At 7^h 45^m, fog not so thick, and some distant trees beyond the hedge just coming in view; temperature of the air 43°; dew-point 42°·5.

At 8^h 30^m, fog very much cleared off; distance distinctly visible, though misty; temperature of air 46°·5; dew-point 44°·5.

At 10^h, fog nearly dissipated, sun partially out; temperature of air 50°; dew-point 48°.

(277.) From the above observations, it is evident

that after a fog has once formed, whatever may be the case when it is first forming, it may continue for a time more or less thick, without a continued saturation of the air with vapour.   The aqueous vesicles, when once precipitated, float about, or are borne along in the breeze, and are only gradually taken up again as the temperature increases, or as they may be wafted into a drier atmosphere.   And in this way there may exist for a time every intermediate amount of visible moisture between the thickest fog and the merest haze only just dimming the transparency of the air.

(278.) I have already mentioned the instance of a thickish fog coming on suddenly one summer's evening, in which the dew-point, at the time of the fog reaching the place of observation, was as much as $6\frac{1}{2}$ degrees below the temperature of the air (270),—thus attesting to a considerable dryness in the air in the middle of the fog.   This, however, tells us nothing as to the exact conditions under which the fog was first formed; and which had been simply borne along by currents to the place in question from a distant region, where the air may have been in a very different hygrometric state. The point which it would be so desirable to ascertain is—the exact relation in which the dew-point stands to other kinds of fogs than those which arise from the lowering of the temperature at the spot in which they first appear (I mean the relation *at the moment of their forming*),—as also what amount of density of fog can coexist, for a time afterwards, with a dew-point still several degrees below the temperature of the air*.

* Humboldt, when surrounded by a thick cloud, at the summit of the Silla of Caraccas, found the air drier than when the sky was serene; and he was much struck with the apparent

(279.) The solving of these questions would throw much light especially upon what have been called *dry fogs*, or fogs that occur principally during very hot dry weather, and which have been thought by some to owe their origin to peat smoke, or to clouds of dust, or to some other foreign bodies, in a state of extreme sub-division, present in the air, though still mixed up with a certain amount of precipitated vapour, and giving to this last that peculiar appearance (along with a peculiar smell), which has sometimes obtained for this kind of fog the name also of *blue mist*.

(280.) Records of these dry fogs, which in certain seasons have been extremely prevalent, and which have extended over vast regions at once, are to be found in most meteorological works. A notable one occurred in the year 1783, which lasted several weeks in the months of June and July, in this country, as well as in several parts of the Continent. Probably, however, in this in-stance, the fog was really occasioned by clouds of dust, having a volcanic origin, as earthquakes and volcanic eruptions are said to have occurred about that time in Iceland, Calabria, Sicily, and other places*. Also the

dryness of the air seeming to increase as the fog augmented. For further remarks by him on this circumstance, see his *Personal Narrative* (Engl. Transl.), vol. iii. p. 518.

* The following is an account of this fog, with some of its attendant circumstances, as narrated by White in his *Natural History of Selborne*:—

"The summer of the year 1783 was an amazing and porten-tous one, and full of horrible phenomena; for, besides the alarm-ing meteors and tremendous thunder-storms that affrighted and distressed the different counties of this kingdom, the peculiar haze, or smoky fog, that prevailed for many weeks in this island, and in every part of Europe, and even beyond its limits, coexistence of similar dry fogs with such phenomena had previously, as it has since, been often observed.

(281.) In some other cases, this kind of fog may be

was a most extraordinary appearance, unlike anything known within the memory of man. By my Journal I find that I had noticed this strange occurrence from June 23 to July 20 inclusive, during which period the wind varied to every quarter, without making any alteration in the air. The sun, at noon, looked as blank as a clouded moon, and shed a rust-coloured ferruginous light on the ground and floors of rooms; but was particularly lurid and blood-coloured at rising and setting. All the time the heat was so intense that butchers' meat could hardly be eaten on the day after it was killed; and the flies swarmed so in the lanes and hedges, that they rendered the horses half frantic, and riding irksome. The country people began to look with a superstitious awe at the red lowering aspect of the sun; and indeed there was reason for the most enlightened person to be apprehensive; for all the while Calabria and part of the Isle of Sicily were torn and convulsed with earthquakes; and about that juncture a volcano sprung out of the sea on the coast of Norway."—*Letter LXV. to Daines Barrington.*

Mrs. Somerville, in her *Physical Geography*, traces the origin of this fog to a volcanic eruption in Iceland. She says, "After the eruption of Skaptar (one of the volcanoes in Iceland), which broke out May 8th, 1783, and continued till August, the sun was hid many days by dense clouds of vapour, which extended to England and Holland."—Vol. i. p. 194. Dr. Daubeny, also, in his work *On Volcanos*, speaking of this same eruption, says, "The atmosphere throughout the whole of the island (Iceland) was obscured, for months after the eruption had taken place, by the enormous quantities of fine dust suspended in it, and traces of the same were perceived even so far off as Holland."—2nd Edit. p. 308.

Thomson, in his *Introduction to Meteorology*, mentions an instance in which volcanic dust was carried a distance of about a thousand miles (p. 160).

caused, not by *volcanic* dust, but by the presence of various other particles, extremely small as well as numerous, of both mineral and animal origin, which always exist to a greater or a less amount in the atmosphere, in hot dry weather, and which would abound the more, in proportion to the length of time the heat and drought had prevailed*.   If, under such circumstances, any precipitation of vapour take place from the intermixing of different currents of air, causing haziness, the density of such a fog would be greatly increased by the presence of the above particles, and be quite disproportionate to its humidity, so as in some measure to warrant the appellation of dry fog.

(282.) I conceive, however, that many of the fogs

* See some remarks on Atmospheric Dust by Professor C. Piazzi Smyth, in an article entitled "Meteorological and Astronomical Notices," published in the *Edinb. New Phil. Journ.* vol. li. p. 381.   He says that "the air is never altogether free from some of these floating particles;" but he gives reasons for believing "that this atmospheric dust is chiefly confined to a height, according to the season of the year, of from 3000 to 6000 feet above the surface of the earth, often preserving a very distinct upper surface to the stratum."

Amongst the particles of organic matter which go to make up this atmospheric dust, are comprised probably, at certain times of the year, immense multitudes of the spores of Fungi, which some have thought may be the source of epidemic diseases, especially of cholera.   See an article by Dr. Daubeny, in the *Edinb. New Phil. Journ.* for July 1855, entitled, "On the Influence of the Lower Vegetable Organisms in the production of Epidemic Diseases."   When heavy rains and thunder-storms occur after much drought and heat, they have the effect of washing to the ground all these extraneous particles in the atmosphere, or, as the expression is, of *clearing the air*; and the thick mistiness immediately passes off.

to which have been given the names of *dry fog* and *blue
mist*, those especially which are of short continuance,
though occasionally much increased by the circumstance
last mentioned, are hardly distinguishable from that
thick haze, which has been already spoken of in a former part of this work (57), as often attending the first
setting-in of hot weather, and which also not unfrequently comes on rather suddenly *during* hot dry
summer weather, being in each case, as I feel assured,
the result of a shift of the wind, generally from Easterly
towards the S. and S.W., but sometimes in the opposite
direction, the shift, however, not always originating in
the current next the earth, and therefore not always
witnessed to immediately by the vane.

(283.) So far as my observations go, previous to these
mists coming on, the sky is always cloudless, though by
no means clear, and the air in a very dry state.   After
the mist has lasted a longer or a shorter time, it either
passes off, without any marked alteration of weather, if
the wind have only shifted a few points, or the effect
of it be very local; or else clouds soon begin to show
themselves, first cirrus and cirrocumulus, the cumulus,
followed perhaps by rain within a day or two after, if
the change of the wind be more decided.

(284.) If, at the time of the wind shifting in this
way, during hot summer weather, clouds already exist,
these latter undergo some modification consequent upon
the change, perhaps become larger and more threatening
or more generally diffused, but no mist occurs like what
has been described above.

(285.) These blue mists, even when due to nothing
more than a mere change of wind after long drought,
sometimes reach over a wide extent of country at the

L 5

same time. Dr. Martin, in his work on the "Undercliff*," has mentioned two or three striking occasions
on which he observed them at Ventnor in the Isle of
Wight; and I find, on looking into my own Meteorological Journals, that the same mists were noticed by
myself at Swaffham Bulbeck, presenting just the character he describes, and occurring during exactly similar
weather, the only difference being, that in one instance
the mist was seen in Cambridgeshire the *day before* that
on which it was observed in the Isle of Wight. In this
last case, was it that the mist took a certain time to
travel over the intervening distance ? or were there local
variations of the wind, by which it was caused at the
two places independently of each other ? This blue mist,
which was one of the most remarkable I ever witnessed,
objects even at a short distance appearing quite indistinct, occurred at Swaffham Bulbeck on the 12th of
June, 1845. It was noticed early in the morning of
that day, and prevailed throughout the forenoon, the
wind being E.S.E., the barometer standing at 10 A.M.
at 30·194 inches, and the thermometer at the same
hour at 72°: in the afternoon the mist gradually cleared
off, cumulus clouds formed, and the temperature rose
to 80°. The weather had been extremely fine, and
without any rain, since the 8th. On the 11th, the day
previous to the mist, the wind was N.E., the barometer
as high as 30·251 inches, the temperature never above
74°, though the sky was bright and cloudless: in the
evening, *a very deep red sunset* led to the suspicion of
some change of the wind being at hand; and the increased heat, with a slightly falling barometer on the

* The "*Undercliff*" of the Isle of Wight; its Climate, History, and Natural Productions, 1849, p. 105.

following day, independent of the mist, was at once an indication of the quarter to which it had shifted during the night. The weather still continued fine for a few days after, but the barometer kept falling, the wind on the 16th worked through the S. to S.W., and on the 17th the fair weather was broken up for a time by rain and thunder-storms.

(286.) I have entered into the above details, because I could clearly trace the origin of the blue mist in this instance, to the first intermixing of two currents of air of different temperatures, after some days of previous drought, the warmer one from the S.E. flowing in upon that from the N.E., which had prevailed the day before the mist came on. I doubt not but in most other instances of mists of a similar character, if all the circumstances of the weather before and after were closely watched, they would be found to arise from the same cause. Dr. Martin, indeed, expresses an opinion in his book, that the blue mists which he observed in the Isle of Wight, and which I have mentioned as identical, in respect of the period of their occurrence, with some noticed by myself in Cambridgeshire, might possibly be due to the *burning of peat*, an operation known to be carried on extensively at certain seasons in Holland and Germany (as it also is, sometimes, in England), to render the soil more fit for agricultural produce, and that the smoke caused by this burning was wafted by the wind from the coast of Holland to our own shores*. But though this peat-burning is often a source of dry fog in the

* For some account of this *peat-burning*, see a paper by M. C. Martins " On the Nature and Origin of different kinds of Dry Fogs," in which this whole subject is treated in some detail, in the *Edinb. New Phil. Journ.*, vol. lvi. p. 229.

countries above mentioned, I feel satisfied that the explanation does not apply here; and setting entirely aside fogs from such sources, as well as those caused by volcanic eruptions, which latter are only of rare occurrence, there can be no doubt that other dry fogs do occasionally show themselves in this country in hot weather, very distinguishable from the *humid fogs* of autumn and winter, and which are only to be accounted for in the manner above alluded to; viz. by a certain amount of precipitated aqueous vapour, temporarily disturbing the transparency of the air, in consequence of a change of wind, under the particular circumstances, and at the season in which these fogs are observed.

(287.) It has been said that "fog cannot possibly form when the air is very dry;" and that "the expression of dry fog contains a contradiction*." No doubt this is to a certain degree true; but what amount of humidity must be present in the air to form fog, or how far it may fall short of actual saturation, and yet allow of a precipitation of vapour, on the first mixing of two currents of different temperatures, is the point which does not appear to have been satisfactorily determined, and upon which further information is so much needed. Assuming, however, that the air cannot be very dry in the exact spot, and at the very moment, in which the fog forms, it must be remembered that, after it has formed, it is often carried by light winds to distant parts, or it gradually descends to a lower region of the atmosphere than that in which it formed, where the air is really dry, comparatively speaking, but where, nevertheless, the fog can continue to exist for a certain time before it is again taken up by evaporation. Observers

* *Kämtz's Meteorology* (by Walker), p. 115.

who speak of it as a dry fog, speak in reference to the
locality in which they are placed, when the fog, which
may have had its origin far off, reaches them. They
are seldom, perhaps, so circumstanced as to be able to
determine the exact hygrometric conditions of that
stratum of air in which the condensation of vapour first
began. Another reason, also, why these fogs are called
dry, is that, though undoubtedly they consist, like other
fogs, of aqueous vesicles, these vesicles, which are probably extremely small, are not deposited upon the person or other bodies with which they come in contact, on
account of these last having a higher temperature than
the fog, whereby there is neither any sensation nor appearance of humidity. In the case of the thick drenching fogs which occur on cold autumn mornings, and
occasionally throughout the winter, not only are the
vesicles of condensed vapour much larger, but they deposit a copious moisture on the surfaces of bodies exposed to them, especially of those which by their nature
are good radiators, and which consequently are cooled
down sometimes much below the temperature of the air.
Every one who has had occasion to be out in one of
these fogs, must have noticed the readiness with which
it is deposited upon the hair and eyelashes, as well
as on woollen garments (which, along with all other
filamentous substances, radiate freely), the vesicles continually coalescing until they collect in large drops.
Sometimes this phenomenon is shown more strikingly
by the condensation of the fog upon the cold leaves and
twigs of the uppermost branches of trees by road-sides,
causing the moisture to run down so copiously, as to
form puddles at the foot of the tree, whilst all the road
lying beyond the drip is quite dusty, if the weather have

been fine and dry previously*. When such very humid
fogs occur in severe weather, it is the freezing of the
condensed moisture on the boughs of trees which causes
rime†.

* See this circumstance, which I have often noticed myself,
alluded to in *White's Natural History of Selborne*, Letter **XXIX**.
to D. Barrington; also in *The Journal of a Naturalist*, 3rd
Edit. p. 59.

† The rime on trees often forms a very beautiful object. The
following is Howard's description of its appearance on the 4th
of January, 1814, one of the severest winters that has hitherto
occurred in the present century. "The mists which have prevailed for several days are probably referable to the modification
stratus. The air has been, in effect, loaded with particles of
freezing water, such as in a higher region would have produced
snow. These attached themselves to all objects, crystallizing
in the most regular and beautiful manner. A blade of grass was
thus converted into a pretty thick *stalagmite*: some of the shrubs,
covered with spreading tufts of crystals, looked as if they were
in blossom; while others, more firmly incrusted, might have
passed for gigantic specimens of white coral. The leaves of
evergreens had a transparent varnish of ice, with an elegant
white fringe. Lofty trees, viewed against the blue sky in the
sunshine, appeared in striking magnificence: the whole face of
nature, in short, was exquisitely dressed out in frost-work.
When the sun, at length, broke through, and loosened the rime,
it fell unmelted, and lay in heaps under the trees; after which
a deep snow, brought by an easterly wind, reduced the whole
scenery to the more ordinary appearances of our winter."—*Climate of London*, vol. ii. p. 225.

Sometimes, in very severe weather, the aqueous particles of
the atmosphere, even when not sufficient to constitute a fog,
become frozen, without first attaching themselves to objects,
and then present a remarkable appearance as they float
loosely about. A case of this kind is mentioned in *White's
Selborne*, as follows, in an account of the frost of December,
1784:—"A circumstance that I must not omit, because it was

(288.) The peculiar smell which has been noticed by many observers as accompanying dry fogs, I conceive to be chiefly due to the precipitated vapour moistening the earthy and other particles so abundant in the air in fine hot weather, and which has been already spoken of under the name of atmospheric dust (281). It is of a similar nature to the smell that occurs after rain, and which is the more striking in proportion to the length of time for which a previous drought may have prevailed. It is also more or less attendant upon all fogs; though when partaking of a sulphurous character, as in the case of some of the dry summer fogs, it is probably connected with electricity, which may likewise have its influence in giving to these fogs some of their other peculiarities, not always capable of an easy explanation*.

new to us, is, that on Friday, December the 10th, being bright sunshine, the air was full of icy *spiculæ*, floating in all directions, like atoms in a sunbeam let into a dark room. We thought them at first particles of the rime falling from very tall hedges; but were soon convinced to the contrary, by making our observations in open places where no rime could reach us. Were they watery particles of the air frozen as they floated; or were they evaporations from the snow frozen as they mounted?"—*Letter LXII. to Daines Barrington.* The former, probably, is the right explanation.

* Forster seems to think that the smell after rain is likewise connected with electricity. He says, "The strong and refreshing smell, which sometimes results when showers first fall after a long drought, is not an invariable attendant on them, even under these circumstances. The highly electrified water of summer's thunder-showers produces this smell the strongest; and it is weakest with the cold, and perhaps, non-electric rain, which sometimes falls after the condensation of a spreading sheet of cirrostratus into nimbus, with a cold atmosphere."— *Researches about Atmosph. Phenomena,* 2nd Edit. p. 226, note.

(289.) Some have referred to the class of dry fogs, the thick fogs which occur in London, and many other large towns, in the colder part of the year; but these are manifestly due to local causes, and have nothing to distinguish them from the regular autumnal and winter fogs of the country, excepting their greater density and less humidity, arising from the smoke of the chimneys, with which they are mixed up. From the cold, and generally at such times, heavy state of the atmosphere, the carbonaceous particles are checked in their ascent, while at the same time their gravity is increased by the moisture they imbibe from the precipitated vapour. That this is the true origin of the thick London fogs, is evident, not merely from the circumstance of their increased density in the morning, soon after the lighting of fires commences, but from that of their so rapidly changing colour from white to yellow, and the contrary (as persons in the west end of the town must have often noticed), when the wind is a little variable, and according as it happens to blow the City smoke either to or from the place of observation *.

---

* White thought that the London smoke was the origin of the " blue mist" which he occasionally observed at Selborne, though he reckons the place about fifty miles distant from the metropolis. He says, " It has somewhat the smell of coal smoke, and as it always comes to us with a N.E. wind, is supposed to come from London. It has a strong smell, and is supposed to occasion blights; when such mists appear they are usually followed by dry weather."—*Naturalists' Calendar* (by Aikin), p. 143. It is more likely, however, that the Selborne blue mists are only instances of the *dry fog* above spoken of, consequent upon the gradual mixing of two currents, one from the N.E., the other from the S.E., and which only occur when the air is in a very dry state.

RAIN.

(290.) The general conditions under which rain falls have been already noticed (233). But it is not easy to assign the exact cause of it in every particular instance. The Huttonian theory of rain, which refers every kind of precipitated vapour to the intermixing of strata of humid air at different temperatures, is hardly sufficient to account for all the phenomena. It is thought by most modern observers that the aid of electricity must be called in for this purpose; and this is the more probable, from the circumstance of the heavy rains which ordinarily accompany thunder-storms, and the increased quantity, in particular, that falls immediately after the electric discharge has taken place.

(291.) Waiving, however, the subject of electrical agency, which is but imperfectly understood, the rains of this climate appear to originate in two distinct ways. Sometimes they are unquestionably due to the collision of opposite currents, as Hutton supposes; but at other times they may be traced to the simple condensation of the vapour in the air by cold, without any intermixing of strata of different temperatures. This last, Howard has observed, "happens when, either the air, flowing constantly from South to North, leaves the influence of the sun behind it; or the sun, declining in autumn and retiring to the Southward, leaves the air to cool where it remains. In effect, both causes may be in action together; as is probably the case during some part of every autumn in these latitudes*." In another part of his work, Howard states, "as matter of experience, that the contact and opposition of different currents *charged with aqueous vapour,* and (by inference from their state

* *Climate of London,* vol. i. p. 117.

as they manifest themselves in succession at the surface) *differing in temperature*, is largely concerned in the production of our vernal and æstival rains." But, " in the decline of the year," when " the great body of the atmosphere is usually moving with some force from S.W. to N.E., while the sun is declining to the Southward," he seems to regard the rain as rather originating in the way above alluded to*.

(292.) Daniell's view of the rains of this climate would seem, in the main, to agree with Howard's, though he does not associate them, according as they take the form of showers, or are of a more lasting character, with different periods of the year. Speaking of the " two sources," to which " the supply of vapour, which occasions rain, may be traced," he says,—" One is the evaporation of the latitude itself, where it is precipitated; and the other, the stream which is perpetually struggling to advance from the equatorial zone. These causes sometimes act conjointly, and sometimes separately. Rain from the first, is derived from sudden falls of temperature, produced by cold currents, or the changes of the seasons, and assumes the form chiefly of showers of greater or less continuance. The expenditure of vapour is but slowly supplied, and the precipitation occurs at intervals. Rain from this source is always accompanied by a declining temperature. When, on the contrary, in consequence of diminished pressure, the tropical current reaches in succession the colder parallels, the supply continues in a perpetual flow; the temperature is raised, the depression of the barometer increases, and rain descends with little intermission †."

* *Climate of London*, vol. i. pp. 127, 128.
† *Meteorological Essays*, p. 116.

(293.) The quantity of rain that falls in different places varies exceedingly ; and, as before observed (197), the humidity of a climate is quite independent of that quantity. In England, as a general rule, the quantity increases from East to West. In some of the midland counties, indeed, according to different returns, the quantity scarcely exceeds what falls in the Eastern ; but on getting further West, it much increases, and in Devon and Cornwall is at least two-thirds in excess of, in some particular localities nearly double, what falls in Cambridgeshire and the neighbourhood of London.

(294.) The quantity of rain is generally much greater in mountainous than in level countries. This, no doubt, is owing to the impediment which the mountain sides present to the free course of horizontal currents charged with vapour, forcing them to ascend to an elevation, at which, from the diminished temperature due to such height, the vapour becomes condensed. The same currents passing over a level country, where they meet with no such resistance, may flow on for a considerable time, without any precipitation taking place*. Hence in the mountainous parts of Cumberland and West-

* See Dr. Hooker's *Flora Indica*, Introductory Essay, p. 76.— The quantity of rain that falls in some parts of India so enormously exceeds what falls in Europe, that, were not the testimony upon which the estimate rests derived from the most unimpeachable sources, it would hardly obtain belief. Dr. Hooker speaks of the climate of Khasia in particular, as remarkable for its excessive rain-fall. "Attention (he says) was first drawn to this by Mr. Yule, who stated, that in the month of August, 1841, 264 inches fell, or 22 feet ; and that during five successive days, 30 inches fell in every 24 hours ! Dr. Thomson and I also recorded 30 inches in one day and night, and during the seven months of our stay, upwards of 500 inches fell, so that the total

moreland we find that even a larger quantity of rain falls than in Devonshire and Cornwall*.

(295.) But it is a remarkable circumstance, that though more rain falls in the vicinity of mountains than at the level of the sea, "the amount at stations *abruptly elevated* above the surface of the earth diminishes as we ascend." This had been long ago suspected to be

annual fall perhaps greatly exceeded 600 inches, or 50 feet, which has been registered in succeeding years! From April, 1849, to April, 1850, 502 inches (42 feet) fell. This unparalleled amount is attributable to the abruptness of the mountains which face the Bay of Bengal, from which they are separated by 200 miles of Jheels and Sunderbunds."—*Himalayan Journals*, vol. ii. p. 282.

* The Lake District has long been known to be the wettest in England. But the quantity of rain that falls in certain localities there, as measured by Mr. Miller, far exceeds anything that had been previously supposed. In a communication made by him to the British Association, he mentions that at Seath-waite, " a small hamlet at the head of the Vale of Borrowdale," the rain that fell in the year 1845 amounted to the astonishing quantity of 151·87 inches. On the 26th and 27th of November alone there fell nearly 10 inches, "being the greatest quantity of rain which has ever been measured in the same period in Great Britain."—(See *Rep. Brit. Assoc.* 1846, Trans. of Sects., p. 18, and *Edinb. New Phil. Journ.* vol. xlii. p. 43.)

In the year 1848, the quantity of rain at Seathwaite is said to have amounted to 160·89 inches! Since then, Mr. Miller has discovered a new locality in the Lake District, which is con-sidered as yielding one-third more rain than even Seathwaite. At this new station, called "the Stye," on Sprinkling Fell, about a mile and a half distant from Seathwaite, the quantity of rain that fell in 1848 was computed to have been not less than 211·62 inches. See Mr. Miller's papers "On the Meteor-ology of the Lake District of Cumberland and Westmoreland," in *Phil. Trans.* 1849, p. 323, and 1851, p. 623.

the case, but it is of late years that, by carefully conducted experiments at York, Professor Phillips has not
only completely established the fact, but been able to
refer it to the right cause. Having placed three rain-
gauges, one on the ground, and the other two at the
respective heights of 44 and 213 feet, he found, on
taking the results of twelve months (1833–34), that
the quantity of rain collected at the three different stations was as follows :—

Height above ground.	Rain in inches.
0 feet....................	25·706
44 „	19·852
213 „	14·953

" The diminution was therefore 41·8 per cent. for
213 feet, and 22·8 per cent. for 44 feet*."

(296.) This difference is considered as due to the
drops of rain enlarging as they descend, in consequence
of the condensation of fresh vapour upon their cold
surfaces, while passing through a humid atmosphere.
Howard had previously suggested another cause for an
increased quantity of rain on the ground than at greater
elevations, which may sometimes combine with the one
just mentioned, and add to the effect. He considers
" that, when rain takes place with a turbid atmosphere,
a considerable and variable proportion of the water is
actually separated from the vaporous medium, at a
height" from the ground which is inconsiderable compared with those elevated regions from which some
showers are precipitated. It is probable, however, that
this occurs principally, if not only, during the winter
half-year, " the lower atmosphere in summer being
generally clear of that misty precipitation, which in the

* Forbes in *Report of Brit. Assoc.* 1840, p. 112.

winter months must contribute something considerable towards the product at the ground*."

(297.) On the same principle that, generally speaking, more rain is received into a gauge on the ground, than at elevations immediately above it, it must sometimes happen that there is less. If the rain descend from a great height, and pass, in falling, through a stratum of air that is relatively dry, the drops will lose somewhat of their size from evaporation before they reach the ground. We occasionally see an extreme case of this phenomenon, before alluded to (246), viz. when a shower, which may be distinctly observed falling in the higher regions of the atmosphere, is *wholly* taken up by an inferior stratum of dry air through which it descends, so that none of it comes to the ground at all.

(298.) But in any case, and without reference to the relative quantities of rain at different heights, the rain-gauge gives but an imperfect return of the exact quantity that falls at a particular spot. Let the gauge be placed how it will, and in the most exposed situations, away from all trees and buildings, still much rain that strikes against the sides of the funnel must, in the case of hard showers, and still more in that of hail, bound out again and be lost†. In the case of snow too, which collects principally in the upper part of the funnel, until it is melted and runs down into the vessel beneath, it is still more difficult to obtain all that falls.

* *Climate of London*, vol. i. pp. 97 and 104.

† Whether there is any loss in the case of showers that fall slantingly, under the influence of a brisk wind, has been differently estimated by different observers. See this question considered in Howard's *Climate of London*, vol. i. p. 98. *He* thinks not.

(299.) My observations on the rain at Swaffham Bulbeck were made with a gauge for the most part similar to that described by Howard, in the Introduction to his work*. The funnel was of brass, with a mouth-piece of six inches diameter. The bottle,—which held at least three quarts, to allow for any unusual fall of rain, when a previous quantity might not have been measured off,—was placed in a box sunk in the earth, so as to bring the mouth of the funnel nearly on a level with the turf. The situation chosen was an open part of a mown lawn some little distance from either trees or houses. Two glass measures were used, both graduated by Newman in reference to the diameter of the funnel : one, capable of containing half an inch of rain, divided into 50 parts, or measuring tenths and hundredths,—the other containing one-tenth of an inch, divided into 100 parts, or measuring thousandths. This last was employed when the quantity of rain to be measured off was very small.

(300.) *Mean quantity of Rain.*—It requires a great number of years' observations in order to get a mean value of the quantity of rain that falls in any particular place. The fall varies exceedingly, not merely from one year to another, but from one series of years to another, if the series be not considerable. This is very observable in some returns of the quantity of rain fallen at Lyndon in Rutlandshire, sent to White by one of his correspondents, and mentioned by him in his *Natural History of Selborne*†. From these returns it appears that the quantity might be estimated at $16\frac{1}{2}$, $18\frac{1}{2}$, $20\frac{1}{4}$, $25\frac{1}{2}$, 26, or even 32 inches, according to the number of years,

* *Climate of London*, vol. i. Introduction, p. xvii.
† Letter V. to Pennant (note).

and the particular years, of which an average was taken, the lowest number of years being 3, and the highest 23. Even when taking a term of ten years, and setting one decade against another, a difference is found of $7\frac{1}{2}$ inches, the mean quantity from 1740 to 1750 being $18\frac{1}{2}$ inches, whilst the mean from 1770 to 1780 is 26 inches.

This is a remarkable discrepancy; and though it has been sometimes thought that there are cycles of wet and dry, as well as of cold and hot seasons, that recur periodically, one is hardly prepared for any variation in the quantity of rain in a given locality to the above amount, between two terms of years so considerable as ten each.

(301.) My own observations on the rain-fall at Swaffham Bulbeck, were not conducted for more than seventeen years. The results are given in the following Tables.

TABLE I.—*Showing the quantity of rain fallen at Swaffham Bulbeck during each of seventeen years, commencing* 1833, *and ending* 1849.

Year.	Rain in Inches.	Year.	Rain in Inches.
1833	18·750	1842	21·598
1834	14·980	1843	27·758
1835	23·310	1844	22·231
1836	23·920	1845	22·652
1837	19·670	1846	22·946
1838	17·390	1847	20·719
1839	26·345	1848	27·711
1840	17·412	1849	23·657
1841	31·220		

(302.) The mean fall of rain at Swaffham Bulbeck, deduced from the observations of the above seventeen years, is 22·486 inches. Allowing for slight occasional losses from evaporation, and other causes, it would probably amount to at least $22\frac{1}{2}$ inches, which may be considered as the mean in round numbers. This differs very little from the mean yearly fall at the Cambridge Observatory, which, according to a return kindly sent me by Professor Challis, is 22·31 inches, calculated from thirteen years' observations, viz. 1838 to 1853 inclusive, omitting 1844, 1845, and 1848, during which years the rain-gauge was out of order.

(303.) Howard sets the mean quantity of rain in the London district, in round numbers, at 25 inches*, which is 2½ inches more than at Swaffham Bulbeck. The mean quantity at Greenwich, on an average of twenty-five years, according to Belville, is nearly similar, being 25·09 inches†. Whether the Cambridgeshire mean would be increased by the observations of a longer period of years, I cannot say; but I find no such difference between the above mean and the mean of any less number of years, exceeding three, at all approaching that above alluded to in the case of the Lyndon observations (300). If we take the mean rain at Swaffham Bulbeck, as deduced from the first ten years in the above series, i. e. commencing with 1833 and ending with 1842, it is found to be 21·459 inches. If we take the mean of the last ten, commencing with 1840 and ending with 1849, it is found to be 23·790 inches, being a greater quantity than the former by 2·331 inches. But neither of these means differs so much as 1½ inch from the mean of the entire seventeen years.

(304.) The wettest year by far in the above series was 1841, when the quantity of rain that fell amounted to 31·220 inches, which is about 3½ inches less than the average fall in some parts of Devonshire, according to returns furnished me by a relative in that

* Vol. i. p. 131.   † *Manual of Barometer*, p. 33.

M

county*. The driest year was 1834, when the quantity
did not exceed 14·980 inches, being actually less than
one-half the above, and showing strikingly how the
amount of rain varies in different years, when only single
years are taken for comparison, at the same place. If
we deduct the lowest mean from the highest, it gives a
difference of no less than 16·24 inches, as the range,
within which the annual quantity of rain varied at Swaffham Bulbeck, during the above seventeen years.

(305.) Howard thinks that about one year in five
attains the maximum of drought, and one in ten the
maximum of wet†. But it is doubtful whether any
such rule would be found to hold good in a long series
of years. The only observable result, with respect to
the recurrence of wet and dry years, in my own observations, is—that all the drier years occur during the
first eight in the series, the quantity of rain in five out
of the eight *not amounting to* 20 inches, whereas in the
last nine, three of the years are the wettest in the whole,
and in all of them the quantity *exceeds* 20 inches.

(306.) There appears to be no constant relation between the yearly mean temperature and the quantity of
rain. Howard, indeed, in one part of his work observes, in reference to a series of years from 1810 to
1816, "that the warm years were uniformly dry, or
below the average in rain, and the cold ones uniformly

* The Rev. W. Heberden of Broadhembury, near Honiton,
who has measured the rain in that locality for a long term of
years, was kind enough to send me the results of twenty years'
observations, commencing with the year 1837. From these
results it appears that the mean yearly fall at Broadhembury is
34·78 inches. The wettest year in the series is 1852, when the
quantity amounted to 49·66 inches; the driest 1854, when it
was reduced to 24·92 inches.   † Vol. i. p. 101.

wet, or above the average*." But this does not appear to be confirmed by the results of a longer series of years, given by himself subsequently†, in which there are many exceptions to such a rule. In two of the wettest years in this last series, the mean temperature was but one or two-tenths of a degree below the average mean, and the next wettest was one of the warmest of all, the mean temperature being above 51°. Neither can I trace any such connexion between the mean temperature and the quantity of rain, in more than a very few instances, in the years over which my own observations extend. At the same time it seems natural that there should be a tendency to such a coincidence, which, when it is not observable, may be counter-influenced by other causes in operation.

TABLE II.—*Showing the average quantity of rain in each month of the year at Swaffham Bulbeck, deduced from seventeen years' observations; also the greatest and least quantity in each month during the same period,—in inches.*

Month.	Mean Fall.	Greatest Fall.	Least Fall.	Range of Fall.
January	1·452	2·313	·360	1·953
February	1·502	3·212	·270	2·942
March	1·633	2·900	·540	2·360
April	1·470	3·765	·210	3·555
May	1·768	4·600	·234	4·366
June	2·035	4·022	·550	3·472
July	2·008	3·720	·680	3·040
August	2·475	5·765	·210	5·555
September	2·164	3·926	·271	3·655
October	2·575	4·383	1·020	3·363
November	1·903	3·645	·921	2·724
December	1·364	2·903	·183	2·720

* Vol. i. p. 101. † *Id.* p. 129.

(307.) Few persons can have failed to notice that some months of the year are generally wetter than others, though there is often a mistake with respect to that one which may be called the wettest. It appears clearly from the above Table, that the wettest month in the year at Swaffham Bulbeck is October, and the driest December. After October the next wettest is August; then follow June, July, September and November, all which months are nearly equal in respect of the quantity of rain that falls in each, while the remaining months in the year are comparatively dry. Thus, in a general way, the year may be divided into a wet half and a dry one, the former comprehending the seasons of summer and autumn, the latter those of winter and spring.

(308.) The exact way in which the rain is distributed through the several months, will of course be subject to variation in different localities. But it will probably be found nearly similar in most of the Eastern and South-Eastern counties of England. On comparing the above Table with a Table of the monthly fall of rain at Greenwich, deduced by Belville from the observations of twenty-five years*, it so far agrees with this latter, as that at Greenwich also October is shown to be the wettest month, while likewise there the year is divisible into six consecutive wet and six consecutive dry months, the chief difference being that, in the Greenwich Table, December falls into the wet half and June into the dry, contrary to what occurs at Swaffham Bulbeck.

(309.) The Table of the quantity of rain in the London district given by Howard†, on an average of thirty-four years, agrees with the Greenwich Table as to the wet and dry months, but makes November the wettest

* *Manual of Barometer*, p. 33.   † Vol. i. p. 136.

month in the year instead of October, and after November, July.

(310.) The driest month in the Swaffham Bulbeck series is December; but as this does not accord with the results of other registers, I think, from some cause or other—perhaps owing to snow, which, when in any quantity, it is difficult to measure accurately,— the quantity for this month may stand below the correct average. However, it is not very much less than the rain of January and April, the latter of which months is accounted the driest by some meteorologists. Howard considers March as the driest. In Belville's Greenwich Table, February is the driest month*.

(311.) The greatest quantity of rain ever registered by me at Swaffham Bulbeck in one month was 5·765 inches. This was in August. The same month shows the greatest variation in the quantity, the range of the fall being 5·555 inches, which exceeds by more than an inch that of any other month of the year. This is due, however, to the accidental circumstance of the exceedingly heavy storm which occurred in this month in 1843, of which more will be said further on. It is the only instance in which the monthly fall of rain amounted to 5 inches, during the seventeen years for which I measured it. A quantity exceeding 4 inches, but less than 5, occurred once in May, once in June, and three times in October.

* According to Dr. Martin, "the wettest months in the Undercliff, in the Isle of Wight, are October and November, after these August; and April is the driest." See his work on *The Undercliff*, p. 95.

At Broadhembury, in Devonshire, according to the returns before alluded to, the wettest month is November, October being nearly as wet, and the driest, March.

(312.) If the mean yearly quantity of rain at Swaffham Bulbeck were distributed equally through the
twelve months, it would give 1·874 inch as the mean
monthly fall, a quantity exceeded, on an average, in
every one of the summer and autumnal months, but
not attained in any of the winter and spring ones.

TABLE III.—*Showing the mean, greatest and least quantity of rain at Swaffham Bulbeck, in each season of
the year, calculated from seventeen years' observations.*

Season.	Mean quantity in inches.	Greatest quantity.	Least quantity.	Range.
Spring	4·871	8·530	1·760	6·770
Summer.....	6·482	11·117	2·740	8·377
Autumn.....	6·813	10·875	3·890	6·985
Winter......	4·216	6·452	2·660	3·792

(313.) From this Table it appears that the rain is
most in excess in autumn, and least in winter: the
quantity in summer, however, falls short of that of
autumn by *less* than half an inch, while the spring
quantity exceeds that of winter by *not much more* than
half an inch*.

(314.) These results must not lead us to a wrong
estimation of the seasons in respect of wet, referring as
they do to *the total quantity* of rain in each season,
without consideration of *the manner in which it falls.*
Though so much less rain falls in winter than in
summer, the number of rainy days, in general, in the

* Dr. Martin makes out the spring (at Ventnor) to be the
driest season, and the summer to be drier than the winter, the
order being as follows:—Autumn, 9·18 inches; Winter, 6·61
inches; Summer, 5·67 inches; Spring, 4·48 inches.—*The Undercliff*, p. 94.

former season, greatly exceeds that of the latter, owing to the rain falling more frequently, but without that violence which attends the heavy storms of June, July, and August. The same amount of rain that takes twelve or twenty-four hours in falling in winter, will sometimes be precipitated during a heavy thunder-storm in summer in the course of an hour or less. The whole quantity, therefore, that falls in winter, though comparatively inconsiderable, will be spread over a much longer period of time than at other seasons, and make the weather throughout more uncertain.

(315.) I have not myself registered, with sufficient exactness to be of much use, the average number of rainy and fine days at Swaffham Bulbeck, in the year or in each month.—" According to Captain Portlock, the average number of dry days (no rain falling) in London, is 220. The number of days of heavy rain varies from sixteen to thirty*."

(316.) Howard says that "in our climate, on an average of years, it rains nearly every other day, more or less," which gives an impression of its great fickleness, such as in truth characterizes it. On a large number of these days, however, the quantity of rain that falls is quite inconsiderable. It must also be remembered that much of the rain falls during the night, the remaining twelve hours, by which persons ordinarily estimate the weather, being left clear.

(317.) Howard, taking separate measurements of the quantity of rain that fell by day, and the quantity by night, for two years, found the latter quantity one-third in excess of the former†; but it may be questioned

* Quoted by Thomson in his *Introduction to Meteorology*, p. 139.
† Vol. i. p. 110.

whether this proportion would hold on the average of a longer term of years. Dr. Martin observes, from ten years' measurement, that, in the Undercliff, in the Isle of Wight, more rain falls during the night than in the day, but that the relative number of rainy days and nights varies with the season*, the nights being oftener rainy than the days in summer and autumn, the days oftener rainy than the nights in spring, while they are nearly equal in winter†.

(318.) *Heavy falls of rain in short periods of time.*— The quantity of rain that falls during the heavy thunder-storms of summer will sometimes exceed the average fall of a whole month. The occasion on which I registered the greatest quantity I ever remember to have fallen at Swaffham Bulbeck in the same period of time, was that of a very heavy thunder-storm on the 8th of August, 1849. This storm only lasted from 5 to 7 P.M.; but during those two hours, the rain, which came down in torrents, mixed with hail of the size of small marbles, and which was measured immediately after the storm was over, amounted to 2·143 inches. As much of the hail, however, that impinged upon the funnel of the gauge bounded out again, the real quantity must have considerably exceeded the above measurement. I had once before, on the 19th of June, 1841, measured upwards of 2 inches of rain after a thunder-storm, but

* *The Undercliff*, p. 97.

† Daniell, also, states, as "the result of experiment, that a greater amount of rain falls while the sun is below, than while it is above the horizon."—*Meteorological Essays*, p. 126. Forbes, on the contrary, speaking of Europe generally, seems to say that more rain falls in the day than in the night, but whether he would apply this rule to England in particular does not appear.—*Report of Brit. Assoc.* 1840, p. 116.

this quantity was not quite equal to the above: the storm, moreover, in this latter instance, lasted several hours longer.

(319.) It rarely rains at Swaffham Bulbeck the whole day, or from sunrise to sunset without stopping. Generally speaking, either the rain does not come on till 10 or 11 A.M., or it commences earlier, and ceases in the afternoon. Exceptions of course occur now and then. One of the most remarkable, so far as my experience goes, was on the last two days of August, 1833, when a storm of wind and rain took place, almost unparalleled, at that period of the year, for its violence and the length of time it lasted. The weather previously had been generally fine from the beginning of the month, with scarcely any rain at all. On the morning of the 30th, a fall commenced from the S. and S.W., at first steady and without wind, but after the lapse of a few hours, accompanied by a furious gale from the N.N.W. (the wind having shifted to this point), which ran tremendously high all that night, and throughout the following day. The rain continued for forty-eight hours, and amounted altogether to more than 2 inches. It was driven with such force by the wind as to find its way into almost every house, and what is more, it quite washed away the plaister from the walls of many of the cottages, laying them bare to the studs and laths in many cases, a circumstance I never witnessed before. The gale also occasioned much damage. Many trees were uprooted,—others snapped in the middle, and carried to a considerable distance. I was told by a respectable farmer, that the stock in the fens were in some places actually blown into the dykes and killed. The storm likewise caused great destruction to poultry, and to the

M 5

smaller birds, which latter were beaten down in considerable numbers, where they could not obtain shelter*.

On Michaelmas Day, 1848, a fall of rain occurred at Swaffham Bulbeck that lasted exactly twelve hours, and amounted to 1·6 inch†.

* Belville has noticed this storm in his *Manual of the Thermometer*, describing it as "one of the heaviest gales on record for the month of August." He says "many trees were blown down in Greenwich Park, buildings were damaged, and the hop-gardens and orchards in Kent sadly injured; the accounts from the coast were appalling, fifty-nine vessels were reported at Lloyd's; in this fearful gale the Amphitrite convict-ship was lost, and all on board perished."—P. 43.

† In a recent communication to the Meteorological Society, Mr. Glaisher has recorded some of the more remarkable falls of rain that have occurred in London during the last twenty-one years. He states that there have been but four instances, in which the fall has equalled or exceeded 2 inches within twenty-four hours. The greatest of these falls took place on the 22nd of October last (1857), when there fell in something more than twenty-four hours 2·75 inches. See *Athenæum*, No. 1570, p. 1489.

Dr. Martin mentions an instance in which 2·80 inches of rain fell at Newport, in the Isle of Wight, during twenty-four hours. —*The Undercliff*, p. 95.

Mr. Miller says that 6½ inches of rain sometimes fall in the Lake District in twenty-four hours, and 10 inches in forty-eight hours.—*Phil. Trans.* 1849, p. 88.

The heavy falls of rain, however, which occasionally take place in this country in short periods of time are as nothing compared with falls which have been recorded on the Continent. Forbes has given some particulars of a fall of no less than 30 inches in twenty-four hours that occurred at Genoa on the 25th of October, 1822. This fact seems well authenticated. The same author has mentioned several other instances of remarkable falls in other places. See *Report of Brit. Assoc.* 1840, p. 113.—See also Thomson's *Introduction to Meteorology*, p. 146, for a long list of instances in different parts of the world.

(320.) The heaviest falls of rain ordinarily occurring (though such as I am about to speak of seldom last above a few minutes) are those caused by a sudden shift of the wind from S. or S.W. to N.W., *during a rain already of some continuance.* Most of the settled wet in this country comes from the S.W.; but it will be generally found that, after the rain from this quarter has continued falling more or less steadily for a few hours, the wind either changes to N.W., or before doing so, passes on to S., the barometer, which is generally low during such weather, sinking lower, and the rain falling more copiously, as the wind approaches that point. The wind seldom remains long in the S. at such times, and though, whilst there, the rain may fall faster than before, it is on the sudden running back of the current next the earth to N.W.,—more especially if, in this backward movement, it jump at once to that point from the S., without stopping in the S.W. a second time,—that the great *splash* of rain, as it may be called, to which I would here draw attention, takes place. I have often known persons, after the occurrence of one of these splashes, describe it as a waterspout, but it does not strictly fall under this class of pheno-mena*. It is evidently due to the sudden mixing of

* Waterspouts and whirlwinds, which are but the same phenomena under different names, are analogous to those gyra-tory movements of the air, which sometimes arise quite suddenly in particular spots, when the air around is generally still, raising and carrying along with them dust, leaves, straws, or any other light substances, which lie in their path. The only difference is the scale upon which these two different phenomena take place.—See *Kämtz's Meteorology* (by Walker), p. 392.

Œrsted defines a waterspout as "a strongly agitated mass of air, which moves over the surface of the earth, and revolves on two airs of very different temperatures, the warmer one from the S. being loaded with vapour, which becomes immediately condensed on coming into contact with the colder one from the N.W. While this momentary deluge is about, the barometer rises, and the air becomes rapidly colder; the latter circumstance being indicated, not unfrequently, by the steam that appears on the inside of windows looking Northwards. But the rain seldom continues to fall with this violence for more than a few minutes; though on one occasion that I noted, the splash lasted for half an hour. After it is over, the sky generally clears, and for a time the rain subsides altogether, though the weather may still preserve an unsettled character.

(321.) As an instance of one of these splashes of rain, I insert the following extract from my Meteorological Journal, under the date of August 13, 1846.

"Wind this morning S.W., blowing fresh, with much mistiness in the air, and clouds forming. Fine, however, till about

---

an axis, of which one extremity is on the earth and the other in a cloud. From this cloud a continuation proceeds downwards, which forms the upper portion of the waterspout; while the lower portion, besides air, consists sometimes of water, sometimes of solid portions, according as the waterspout passes over land or over water."—See his paper on "Waterspouts" in the *Edinb. New Phil. Journ.* 1839, vol. xxvii. p. 52. In another part of this paper, he observes that "waterspouts take place for the most part in still weather, and during unsteady winds." And again, that "they take place most frequently in the midst of a serene atmosphere."

I never saw a true waterspout in Cambridgeshire; and believe that all those reported to have occurred were nothing more than exceedingly hard falls of rain, under circumstances similar to those described above.

2 P.M., when a driving rain commenced, which kept on and off for two hours, the wind latterly working on to S.   At 4 P.M. the lower current shifted suddenly, and with much violence, to N.W., causing the rain to descend in torrents for about a quarter of an hour, *so thick as to render objects out of doors almost invisible.* After this burst of rain had subsided, the air cleared, and it became very fine, the wind continuing N.W. for the rest of the day.   The barometer, which stood at 29·75 at 10 A.M., continued falling till the sudden shift of the wind took place, when it immediately began to rise again.   The temperature of the air, which had been 70°·5 when the rain first came on, was lowered to 58° after it was over."

In another instance (August 2, 1847), after a sudden shift of the wind from S.W. to N.W., accompanied by a heavy storm, the temperature, which previous to the shift had been as high as 81°, fell soon after to 65°, and by 10 P.M. had fallen to 56°·5 !

(322.) It is not easy to estimate the exact quantity of rain falling in one of these splashes, as it comes on quite suddenly *during rain* that may have been already falling for some hours.   It is hardly possible, therefore, to separate it from the *whole* quantity.   Judging, however, from the whole quantity measured on different occasions, in connexion with the whole time for which the rain had lasted, I conceive that, in some instances, it must have come down, during the splash, at the rate of from $1\frac{1}{2}$ inch to nearly 2 inches within the hour.

(323.) Storms and showers are more or less partial, though some of long duration pass successively over a number of places lying in the line of their course, before they are wholly spent.   It is otherwise with rains that come on more gradually, and which last the greater part of the day : these often reach over a very considerable area, falling everywhere within that area at

the same time. Howard mentions an instance in which
he took some pains to ascertain the exact extent of
country over which a steady rain, that commenced early
in the morning and continued till after sunset, had
fallen. He says,—" The rain of the 4th of the tenth
month (October 1813) having put a period to a fair
season of some weeks' continuance, I availed myself of
a journey I made immediately after it, to endeavour to
ascertain by inquiry how far it had extended. I found
it had rained on that day from morning to night through
the whole distance from London to York; as likewise
further North, quite to the Tyne, and across the island
from Cheshire to Northumberland. It having been
likewise a very wet day on the south coast, the conclusion seems to be fair, that the whole of England, at
least, was on this occasion irrigated at once, by the
introduction of a current from the Atlantic, which now
displaced completely the Easterly breeze previously de-
flected from the N.E. to the S.E.; both currents on
this occasion probably depositing the excess of water
with which they were charged*."

(324.) *Transparency of the air before rain.*—It is a
common observation, that before rain there is a peculiar
transparency of the air, which allows of objects being
seen clearly, which are ordinarily indistinct. Fishermen, and other persons who are much abroad, often
learn practically to prognosticate the approach of wet,
by such a circumstance. Their judgment is determined
by the sight of some beacon, headland, or mountain, to

---

* *Climate of London*, vol. ii. p. 217. Thomson mentions that
"on February 3, 1842, it rained in North America, from N. to
S. over 1400 miles: the breadth and boundaries of the district
were not ascertained."—*Introduction to Meteorology*, p. 140.

which they are in the habit of looking on such occasions. This state of the air is directly opposed to that haziness, spoken of in a former part of this work (57, 282), which seems to pervade the whole atmosphere in dry hot weather.

(325.) But the same transparency of the air occurs *after* rain, the cause of which is probably to be found in the drops of rain having beaten down, and cleared the atmosphere of those small particles of dust, &c. which are ordinarily floating in it, and to which allusion has been before made (281). Neither is it a fact that before *settled* rains, especially when they follow a long run of fine weather, the air *is* always clear in the manner asserted. Generally speaking, for some time previous, there is a prevalence of misty cloudiness, which gradually increases, rendering the air more and more turbid, until the rain actually begins to descend.

(326.) The truth seems rather to be that it is during *changeable* weather, when there is a frequent succession of storms and showers, and *between the showers*, that the atmosphere is characterized by great transparency. The dust, and other foreign matter in it, after being beaten down by one storm, has not time to collect again before another occurs. In general, too, when the air is thick, it is not due entirely to such dust, but to a certain amount of precipitated vapour, such as occurs in ordinary mists, mixed up with it. But in showery weather, this precipitated vapour is mostly collected into the large masses of cloud,—chiefly cumulostratus passing into nimbus,—from whence the showers come; instead of being diffused through the air more uniformly, as in fine settled weather, when no clouds exist, or only a few at a great altitude *.

* Thus Howard states that "in unsettled weather the rapid

(327.) Hence, too, this transparency of the air can hardly arise, as is often supposed, from an excess of vapour in the air, *i. e.* of vapour in the aëriform state as it exists previous to precipitation. For first, as a late writer on Meteorology has observed, " it does not appear that the gases (of which the atmosphere is constituted) are more transparent when vapour is present than when it is absent ; nor is there any reason to presume that vapour is more transparent than the gases *." Secondly, if the transparency were due to such cause, however we might expect to notice it previously to the rain coming on, we could hardly explain its prevailing to the same degree after the rain has fallen, and when a certain amount of the vapour, to which it is thought to owe its origin, must have been precipitated †.

formation of large cumuli has been observed to clear the sky of a considerable hazy whiteness ; which on the other hand has been found to ensue upon their *dispersion*."—*Climate of London*, vol. i., Introduction, p. lxiv.

    * Hopkins *On Atmospheric Changes*, p. 367.

    † Humboldt has somewhere made the following observation, though I am unable to give the exact reference :—" The Peak of Teneriffe is seldom seen at a great distance in the warm and dry months of July and August,—and on the contrary is seen at very extraordinary distances in the months of January and February, when the sky is slightly covered, and immediately after a heavy rain, or a few hours before it falls. It appears, that the transparency of the air is prodigiously increased when a certain quantity of water is uniformly diffused through the atmosphere."

    If by "water" Humboldt means vapour in the aëriform state, his opinion is at variance with that of Hopkins above stated, and it is not for me to reconcile the conflicting authorities. Yet he can hardly mean precipitated vapour, or floating aqueous

(328.) Whatever be the cause of this transparency of the air, there is no doubt that, if combined, as it almost always is, with a low barometer, showing no disposition to rise, however fine it may appear overhead, the weather is not to be depended upon.

(329.) The *blue tint of the atmosphere,* which is so far connected with its transparency, as that the sky is never blue when the air is thick from precipitated vapour, will likewise assist us, sometimes, in forming a judgement of the weather. The richest blue is most obser-

---

vesicles like those of mist, which under no circumstances could fail to render the atmosphere *thick* rather than transparent.

That a very dry state of the atmosphere is sometimes combined with great transparency, and therefore not always due to the "quantity of water diffused through it," appears from an observation by Mr. Darwin in the very same part of the world to which Humboldt alludes. Speaking of the atmosphere at St. Jago, one of the Cape de Verde Islands, and after stating that "it is generally very hazy, due," as he believes, "to an impalpable dust, which is constantly falling, even on vessels far out at sea," and thought to be derived "from the wear and tear of volcanic rocks," Mr. Darwin adds the following remark:— "One morning the view was singularly clear; the distant mountains being projected with the sharpest outline, on a heavy bank of dark blue clouds. Judging from the appearance, and from similar cases in England, I supposed that the air was saturated with moisture. The fact, however, turned out quite the contrary. The hygrometer gave a difference of 29·6 degrees, between the temperature of the air, and the point at which dew was precipitated. This difference was nearly double that which I had observed on the previous mornings. This unusual degree of atmospheric dryness was accompanied by continual flashes of lightning. Is it not an uncommon case, thus to find a remarkable degree of aërial transparency with such a state of weather?" —*Journal,* &c. p. 4.

vable, in this climate, between showers, in changeable weather, such as often occurs in the months of April and May,—a period of the year, when, owing to the dry Easterly winds, the air is at all times more free from vapour than at other seasons.

(330.) At considerable elevations, when the observer gets above the greater part of the vesicular vapour, and other opake matter floating in the air, the blueness of the sky assumes a depth of tint approaching to black. In like manner, the blue in the neighbourhood of the horizon, seen through a greater body of vapour, is less intense than at the zenith*.

(331.) *Rain without clouds.*—Rain sometimes falls from a cloudless sky, and when it is quite blue at the zenith; but though the drops may be large on these occasions, the quantity is seldom considerable. Ordinarily the fall lasts but a very few minutes. It has been explained by supposing that, under certain con-

---

* According to Dr. Traill, the azure tint of the atmosphere "depends upon the greater refrangibility of the blue rays of light than of those towards the other end of the spectrum. When a ray of white or undecomposed light enters our atmosphere, the red and yellow rays pass with little deviation from a rectilineal course; but the blue rays are dispersed in the air, and affect the colour of objects beheld through a long column of the atmosphere. As we ascend lofty mountains, the refractive power of the air diminishes with its density, and hence the blue tint is lost in the blackness produced by the want of refraction; until a traveller, on the ridges of the Alps, the Andes, or the Himalaya, will perceive the sky to assume a hue approaching to deep blue or black."—*PhysicalGeography*, p. 159. But several other explanations have been given of this blue tint, as well as of other remarkable colours occasionally exhibited by the sky and clouds.—See Forbes in *Report of Brit. Assoc.* 1840, p. 120.

ditions, the vapour in the air, which is probably in great excess on such occasions, is at once " condensed into water without passing through the intermediate state of *vesicular* vapour*."

(332.) A similar occurrence of large drops of rain falling for a short time from a sky, not entirely cloudless, but when no clouds are visible, excepting a very few inconsiderable patches, either of cirrocumulus or scud, may be often noticed between showers, or after the breaking-up of some heavy storm. If no storm occur at the place of observation, such drops are indicative of one elsewhere. Thus on the 5th of July, 1843, a fine and very hot summer's day, drops of rain fell at Swaffham Bulbeck at intervals during the forenoon, though only a few very small fleecy clouds passed overhead : in the afternoon, the sky was overcast for about an hour, without any rain falling ; it then cleared, but much scud floated along from time to time, at a low elevation, all the evening. It appeared afterwards from the papers that an exceedingly heavy thunder-storm had passed over the country from S. to N., arriving at Nottingham and Derby about 4 o'clock, and thence passing on to different parts of Yorkshire, Lancashire, and the Lake District†.

* Kämtz's *Meteorology*, p. 132. See also some remarks on this phenomenon by Arago, who offers a different conjecture as to the cause.—*Edinb. New Phil. Journ.* vol. xxi. p. 26.

† Some account of this storm, and the direction it took, will be found in Lowe's *Atmospheric Phenomena*, p. 213. It is there stated to have been " one of the most remarkable thunder-storms perhaps on record ; having traversed a space of 450 miles in length, and 150 miles in breadth, and this in the short period of five or six hours."

(333.) From the circumstance of these large drops
of rain, without any obvious clouds, occurring most
frequently in hot sultry weather when storms are about,
they have sometimes obtained the name of *heat-drops.*
They are not always attended by a low barometer, though
the dew-point is generally high: I have known the
former above 30 inches at such times in some instances.

(334.) *Rain in connexion with the different winds.*—
It has been shown in a former part of this work (113,
127), that, notwithstanding the variable character of
the winds in this climate, there are still certain
winds, which, on the whole, prevail in different months
and seasons more than others. It has been further
shown that there is, on the average, a dry and a wet
period of the year (307). Hence it seems possible to
trace a connexion between the rain and the quarter from
which the wind blows. The driest period of the year
is spring, during the prevalence of a wind from some
point between N. and E.; the wettest is autumn, when
the wind is generally between S. and W. But the
summer is also a wet season, often not much less so
than autumn, and during the former season the wind
averages from some point between N. and W. A different class of winds, therefore, is connected with the
rains of summer from that which is connected with
those of autumn. This circumstance, however, is in
exact keeping with the different conditions under which,
for the most part, the rains of these two seasons respectively occur. Mention has been already made of the
two distinct causes to which the rains of these latitudes
are due (291),—one being the collision of opposite
currents of different temperatures, the other the condensation of the great body of vapour, which is always

endeavouring to make its way northwards from the tropical regions. Now it is from the former source that our summer rains are principally derived, while those of autumn have their origin in the latter. If the N.Easterly winds of spring always prevailed, there would be no rain at all; for these winds, passing over vast tracts of land, are naturally dry, and what vapour they contain is further and further removed from the point of condensation, as they gradually arrive in latitudes of a higher temperature. On the other hand, "if the S.Westerly winds blew without ceasing, it would always rain; for, as soon as the moist air gets cool, the vapour of water is precipitated*." When, again, these winds interchange, or give place to a wind temporarily blowing from some intermediate point, we might expect an alternation of dry and wet, the rain falling principally in the form of storms and showers. Now it is just this changeable weather which characterizes an English summer, when judged by the average of a long term of years. The wind in that season, if not more variable than in the spring and autumn, is less confined either to the N.E. or S.W. Its right place, so to speak, or the quarter from which it ordinarily blows, is some point intermediate between those two quarters. The degree to which it inclines towards, or deviates from, either of them, is determined by the relative advantage which one of these two winds has over the other. And it is their frequent collision and intermixture at that season,—when the S.W. current, which brings up the vapour (or the upper trade), reaches the earth in the higher latitudes,—that causes most of the summer storms in our climate. As the

* Kämtz (*Meteorology*), p. 137.

autumn draws on, this current arrives continually in lower latitudes*, leaving the vapour which it had imported in summer to be condensed by cold, whereby arise the more settled rains that occur in the latter part of the year.

(335.) Neither in the instance of the summer rains with a N.W. wind, nor in those instances in which it rains with the wind due E. or N.E.,—which it not unfrequently does, in winter especially,—must we consider these winds as bringing up the rain, but, in the first case, as simply condensing the vapour of some warmer current opposing them in their course,—and in the second, as exercising the same cooling influence upon a generally humid atmosphere into which they flow.

HAIL.

(336.) Hail is generally "regarded as consisting of drops of rain, more or less suddenly frozen by exposure to a temperature below 32°. If the degree of cold has been very sudden and intense, which is often the case, the icy nucleus, from its being of a temperature far below the freezing-point, acquires magnitude as it descends, by condensing on its surface the vapour of the lower regions of the atmosphere. Hence, even under ordinary circumstances, hailstones often become of considerable size, are nearly always more or less rounded, and when broken, are seen to be composed of concentric layers†."

(337.) But the theory of frozen drops of rain will hardly apply to cases in which large masses of ice have fallen to the ground during severe hail-storms, if credit

* Kämtz, p. 145.
† Prout (*Bridgewater Treatise*), 2nd Edit. p. 331.

is to be given to the several statements on record of such phenomena,—or even to those cases in which the pieces of ice, though not so very considerable in size, are still shapeless, and presenting no uniformity of character.   In the accounts given of many heavy storms that have occurred even in our own country, the stones have been described as more " like rugged pieces of ice " than resembling what we usually understand by the term hailstone*.   In such cases, it seems not unlikely that the stones, when falling very thick, are brought by collision into close contact, so as to become agglutinated, whilst in the act of forming, and before the congealing process has advanced so far as to prevent their uniting.   Thus Howard mentions a fearful hail-storm in America, in which many of the stones were " thirteen, fourteen, and even fifteen inches in circumference, the large pieces appearing to be aggregated of numerous others, which were likewise composed of smaller ones†."

(338.) It has been thought probable that in the formation of hail, as in that of rain, electricity plays an important part; and, undoubtedly, the largest hailstones that fall, in this country at least, are attendant

---

* Howard, speaking of some very large hailstones, which fell during an exceedingly heavy thunder-storm in Gloucestershire, July 15th, 1808, says,—" It may be doubted, however, whether such a name be applicable; for the masses of ice which fell on the places where the tempest most fiercely raged, bore no resemblance to hailstones in magnitude or formation, most of them being of a very irregular shape, broad, flat, and ragged, and many measuring from 3 to 9 inches in circumference."—*Climate of London*, vol. ii. p. 50.

† *Climate of London*, vol. ii. p. 257.

upon heavy thunder-storms. Nevertheless the exact connexion between these two phenomena has never been satisfactorily made out, and some are disposed to doubt that any such connexion exists. Professor Olmsted of America would rather assign as the cause of hail-storms, " the congelation of the watery vapour of a body of warm and humid air, by its suddenly mixing with an exceedingly cold wind, in the higher regions of the atmosphere*." Excepting the difference of tem-

* See a paper by him "On the Phenomena and Causes of Hail-storms" (*Edinb. New Phil. Journ.*), 1830, vol. ix. p. 244. See also an article in the same Journal (1836, vol. xxi. p. 280), by M. De la Rive, "On the Formation of Hail," in which the author seems to consider that, though "electricity always accompanies the formation and fall of hail, these two phenomena are not connected as cause and effect," but that they have a common origin in altered distributions of the temperature of the air, from below upwards, through the intervention of clouds. M. De la Rive regards his views as confirmed by observations made by M. Lecoq, in the very regions in which hail is formed, among the mountains of Auvergne. Interesting details are given of the circumstances attending the coming-on of a terrible storm, of which M. Lecoq was a close spectator, especially as regards the appearance of the clouds, and the structure of the hailstones, which latter were mostly about the size of a pullet's egg, but some as large as a turkey's. M. Lecoq observed that there were two strata of clouds, placed the one over the other, and two winds from different quarters, both which conditions he considers necessary for the production of hail. The exact particulars he has narrated as follows :—

"On the 28th of July (1835) the sun rose from an azure sky, no cloud appeared on the horizon, no vapour floated in the atmosphere, so that a beautiful day was anticipated. At 10 A.M. the heat became intense, and at midday it was almost intolerable, and then some thin flakes of vapour floated in the air at a great distance; the wind was north, but so feeble, that it in no

perature, it is under these same conditions that the
heaviest falls of rain likewise occur, and there seems no

degree tempered the heat. At one o'clock the wind had increased ; the white and floating clouds had descended considerably, and half an hour later covered a great part of the horizon ;
they had a greyish tint, which became darker and darker, till
they were nearly quite black. At two o'clock they formed an
immense covering over the whole of Auvergne; and it was
then easy to anticipate that a frightful storm was at hand. We
waited with anxiety for the issue of that majestic and terrible
scene which was preparing. ᴗSilence and consternation everywhere reigned, speedily flashes of lightning illuminated the
massive vapours which covered the old volcanos of Auvergne,
while the sun still shone upon a portion of La Limagne. We
then heard a distant and low-muttering sound, which resembled
a kind of rolling, and almost about the same time we saw a vast
cloud advance from the west to the east, pure white in some
places, but principally on its edges, and of a deep grey colour
in the centre; it approached with great rapidity, and seemed to
be hurried forward by a violent west wind, which we had not
previously felt at Clermont. This cloud was evidently underneath all the others; its borders were festooned and deeply
slashed, and protuberances, in the shape of long nipples, were
suspended from the lower portion. At a quarter past two, the
anterior part of this cloud had approached very near to Clermont, and the noise which we had long indistinctly heard was
now very intense; and I then very clearly distinguished a very
rapid motion in the edges of the cloud; these edges seemed to
me to be undulating, but in the position in which I was, what
appeared to be undulations must have been the product of a
very violent agitation. I then imagined that I could distinctly
perceive hailstones in the edges of the cloud, and I predicted to
some persons who were with me the immediate descent of hail.
Accordingly, two minutes after having seen this whirlwind kind
of motion, there was a fall of hailstones, which instantly broke
all the tiles of the houses, and all the panes of glass exposed to
the north and west; for the hailstones being at the same time

N

reason why rain and hail should not so far have a common origin.   It is observable that they often pass one
into the other, either the hail being followed by rain,
or the contrary.

(339.) A confused noise is heard at the approach of
hail, and, when heavy, for some little time before it
reaches the place where the observer is.   Some think
this noise is caused by the hailstones being driven by
the wind one against the other.   Others attribute it "to
the combination of the individual sounds produced by
each hailstone cutting the air with great swiftness."

(340.) Hail in this country generally occurs in
spring and summer, accompanying the showers and

propelled along both by the north and the west wind, necessarily took the mean direction.   The first hailstones which fell
succeeded each other very slowly, then all at once their number
increased so rapidly that in ten minutes the soil was covered
with them; some drops of water escaped at the same time from
the electrical cloud, and then the distant rolling sound which
we had so long heard entirely ceased; and the cloud, freed from
its swelling appendages, was carried away by the wind; after
some hours the sun illuminated, with its pale and feeble light,
that scene of desolation which night was speedily about to
envelope."

M. Lecoq witnessed a few days after, in the same locality, the
formation of another hail-storm, many of the appearances of
which were similar to those described above.   There were the
same two strata of clouds, the lowermost having indented
edges, and exhibiting in these edges the same whirlwind kind
of movement, as in the first instance, previous to the descent
of the hailstones.   The reader is referred to the paper itself
for further details, as well as for M. De la Rive's remarks on
these two strata of clouds, in connexion with his view of the
common origin of hail, and of the electricity which is always
more or less developed during hail-storms.

storms that are so characteristic of those seasons. The heaviest falls of hail are always during hot weather, and usually during the hottest part of the day ; though I have occasionally known them at night. Hail in winter is less frequent, but still by no means of rare occurrence ; what falls, however, at this season seldom consists of anything more than opake grains, in size scarcely exceeding a pin's head, or a very small pea.

(341.) Except on the occasion of the very heavy thunder-storm in August 1843, spoken of further on, I have never known any great falls of hail at Swaffham Bulbeck, sufficient to do any serious or extensive damage. In that instance the largest stones measured at least 1½ inch in diameter* : they were not all of the same form, some being rounded or elliptic, others of a very irregular shape. In general stones falling from the same cloud preserve a considerable uniformity of character.

(342.) During a succession of hail-storms at Swaffham Bulbeck on the 24th of March 1828, one occurred

* The hailstones that fall in England are seldom of any great size compared with those which fall in other countries. The South of France, in particular, has long been known as the seat of very destructive hail-storms. A Table of some authentic instances of hailstones of unusual weight and dimensions, with their several dates and localities, will be found in the *Encyclopædia Metropolitana* (Art. "Meteorology"), p. 131. A long list of such instances is also given by Thomson (*Introduction to Meteorology*, pp. 178–182).

India also seems to be subject to very violent hail-storms, notwithstanding it has been sometimes stated that they do not occur in tropical countries. See a communication to the British Association by Dr. Buist, " On Remarkable Hail-storms in India," *Rep. Brit. Assoc.* 1855 (Trans. of Sects.), p. 31.

in which the stones were perfect cones, measuring more than half an inch from the base to the apex. A section parallel to the base exhibited a radiated structure. The temperature of the air at the exact time was 40°*.

SNOW.

(343.) Snow is caused by the freezing of the minute aqueous particles of the atmosphere. Though the flakes may originate in a cloud, they probably, in many instances, acquire a considerable increase of size, by the addition of other particles, as they pass through successive strata of cold and humid air in their descent to the earth. Unlike hail, which falls most frequently, and with the greatest violence, during hot weather, snow only falls when the general temperature of the air is at or near the freezing-point. We seldom, however, in this country, so far as my observation goes, have snow in any great quantity, if the temperature be very low indeed : or if it has been low, it rises, sometimes very suddenly, up to near 32°, previous to the snow coming down. The reason is obvious : the coldest weather is always accompanied by a wind varying from E. to N.E., which is the driest quarter, and the air is generally at such times clear and free from vapour. If it snows, as it often does during frost, notwithstanding the wind, as shown by the vane, is in one of the above two points, it is due to an upper S.W. current, which supplies the moisture, this last being congealed by the

* Howard has mentioned an instance somewhat similar, in which he says " the balls were opake, in the form of a cone with a rounded base, about half an inch in diameter, and composed entirely of striæ meeting at the apex of the cone."—*Climate of London*, vol. ii. p. 325.

colder air below.   Such a snow may continue a long
time, but the same circumstance of the S.W. current
above, which gives rise to the snow, tends to moderate
the cold, and to bring the thermometer nearly up to
32°, before the snow actually begins to fall.   After the
snow is all down, either the whole body of the atmosphere works round to the S.W., and a thaw ensues;
or else the upper current falls into the same direction
as the lower, and the wind continues Easterly, the temperature falling lower than it had been at first.   This
severer cold is due to the air being cleared of vapour
by the fall of snow, and thus increasing the effect of
terrestrial radiation at night.   Hence the common remark that snow brings either a thaw or more intense
frost.

(344.) Deep drifting snows are not very common in
Cambridgeshire, partly from the circumstance of the
generally level country affording few hollows in which
the snow can collect,—and partly from that of the snow
seldom continuing to fall for a sufficient time to accumulate to any great degree.   During twenty years I
only remember two or three instances, in which the
roads have been so obstructed with snow as to interfere
with the traffic.   The most memorable of these was
that which occurred in December 1836; and it is observable that this tremendous snow-storm happened
under the exact circumstances described above.   The
snow commenced falling on the morning of Christmas
Day, and continued, with scarcely any intermission, all
that day and the following night, as well as the greater
part of the following day.   The wind was N.E., blowing almost a gale the whole time, and drifting the
snow in some places to the depth of ten and even

twenty feet. The barometer stood at 29·856 inches,
when the snow first came on: it kept falling, though
but slightly, for the next thirty-six hours, until, on the
evening of the 26th, it stood at 29·658 inches, the
snow coming down more thickly, and the wind blowing
stronger, as the mercury lowered. On the 27th, the
barometer began to rise again, and continued rising
until, on the 31st, it had got up to 30·503 inches.
During these five days, the sky was clouded, and snow
still fell at intervals, though in no great quantity. On
the 1st of January the weather was quite fine and settled,
the sky clear, the wind still N.E., from which point it
had never deviated, the barometer as high as 30·59
inches, and the air very cold, the temperature falling at
night to 24°, and on the night following to 18°. During
the time of the worst part of the storm, the temperature
varied from 28° to 32°.

(345.) There can be little doubt that in the instance
just mentioned there was a S.W. current above, which,
mixing with the colder air from the N.E. below, caused
the storm, being indicated by a fall of the barometer to
about the height at which it would ordinarily stand in
fine weather, if the whole or greater part of the atmosphere were moving from the S.W., instead of merely
an upper current. As this current gave way, the storm
abated, and the barometer gradually rose again, until
the elevation of this last to above 30½ inches, and a
clear sky, testified to the N.E. wind having regained
the supremacy.

(346.) The above snow-storm was said in the papers
to have extended over every part of the kingdom. According to the 'Cambridge Chronicle,' the roads from
Cambridge to London, and the North, and indeed to

every other place, were completely stopped. No coaches were able to pursue their journey: some that left the town of Cambridge on the morning of the 26th*, being unable to proceed, were compelled to return.

(347.) It is very difficult to say what the exact quantity of snow which fell during this storm was, as measured in the usual way. The fall continued unintermittently for so long a time, that my rain-gauge was completely buried, and consequently of little use. I suppose that nearly two-thirds of the quantity of snow that ought to have been received into it never entered the funnel at all. So much as was secured, when melted, yielded about three-quarters of an inch of water. If we suppose twice as much again to have fallen, it would amount to two inches and a quarter, which is equivalent to $22\frac{1}{2}$ inches in depth of snow†, and this is about what the snow measured in places where it was not particularly drifted.

(348.) The flakes of snow, which are so extremely beautiful, and present so many different forms‡, are likewise very variable in size. The largest flakes, however, are not those, generally speaking, which cause the deepest snows. Most of the drifting snows that I have noticed have consisted of very fine crystalline particles, borne along in the wind like smoke; and it is perhaps due to the agitation of the air, by which such

* The 25th was on a Sunday that year.

† "Snow is considered to yield one-tenth of an inch of water for every inch of depth: thus, if the snow, when melted and measured, yields one inch of water, it is concluded that the fall was ten inches in depth."—Drew (*Practical Meteorology*), p. 189.

‡ For descriptions and accurate figures of the different forms

storms are generally accompanied, that other particles
are not allowed to aggregate about them, in the manner
before spoken of (343), to increase the size of the flakes,
in their passage to the ground. The largest flakes are
those which occur in the snow-showers, as they may be
termed, of March and April, falling rather suddenly,
and at first slowly, through a calm atmosphere, though
afterwards more thickly, the wind at the same time
rising, as the cloud from whence they descend comes
more completely over the zenith. Such flakes are often
nearly an inch in diameter*, and whiten the ground in
a very few minutes, or even seconds sometimes; but the
fall seldom lasts long,—after a quarter of an hour or
less, the cloud passes by, and the sky is again clear.
There is an analogy here to the difference observable
between the sudden storms of rain in summer, passing
off after a short time, and the more settled rains at

of snow, see *Scoresby's Arctic Regions,* vol. i. The author of
that work arranges them under the five following groups:—

" 1. Lamellar.

" 2. A lamellar or spherical nucleus, with spinous ramifica-
tions in different places.

" 3. Fine spiculæ, or six-sided prisms.

" 4. Hexagonal pyramids.

" 5. Spiculæ, having one or both extremities affixed to the
centre of a lamellar crystal."

The above groups admit of subdivision, the varieties in some
cases being extremely numerous. Several of the most beautiful,
however, occur only at very low temperatures, and appear to be
confined to the Arctic Regions.

Scoresby's figures will be found reproduced in the *Encyclopædia Metropolitana* (Art. "Meteorology"), pl. 3.

* Howard mentions an instance in which the flakes were
about an inch and a half in diameter.—*Climate of London,*
vol. ii. p. 325.

other periods; in the former case the rain falls very heavily and in large drops; in the latter, the drops are individually smaller, and come with less force, but the rain continues, it may be, for a whole day or more.

(349.) But the snow-storms that come on so suddenly in the spring months occasionally last for a longer time than above stated; and a remarkable circumstance, as it appeared to me, accompanied one of these heavy storms, which occurred, during changeable weather, on the 2nd of March, 1827. In this instance, with the barometer standing at 29·33 inches, and the wind blowing very hard from the N.W., the snow commenced at 10 A.M., and continued for four hours without intermission,—falling all this time so thickly, *as to have the effect of congealing the surface of water*, wherever it was sufficiently exposed to receive the flakes. The canal in Bottisham Park, which varied perhaps in depth from two to four feet, was thinly frozen over throughout its whole extent; and one or two smaller ponds acquired a much thicker layer of ice and snow that was not dissipated several hours after. The explanation of this phenomenon, which I never witnessed before, would seem to be—that the snow fell faster than the water could dissolve it; and though of course the surface water, when cooled by the snow, must have sunk to the bottom to have its place supplied by a warmer stratum, the snow continued falling so long, that, after a time, the whole had its temperature sufficiently lowered to produce congelation. It must be observed that there had been no previous frost: the temperature of the air, when the snow commenced, was 42°, and it did not fall below 34°, during the whole time the snow was falling.

(350.) Sometimes, during snow-storms, another curious phenomenon takes place, to which Howard has given the name of *natural snow-balls*; the snow, "instead of driving loose before the wind," being collected into balls which roll on, increasing in size until their weight stops them. Howard speaks of thousands of these balls, as seen on one occasion "lying in the fields, some of them several inches in diameter*." I have once or twice witnessed something similar to this myself.

(351.) Though snow is formed in a region of the atmosphere which is below the temperature of the freezing-point, it often happens that, in its descent, it has to traverse other strata of air, so much warmer, as to arrive at the ground in a partially melted state: it is then called *sleet*. This is frequently the case at the first coming-on of a thaw after a frost, or when the season is in a transition state between winter and spring. As the air too is always colder the higher we ascend, the same precipitated vapour which covers the tops of the hills with a cap of snow, in many instances reaches the plains, at a lower elevation, in the form of rain. The early and late appearance of snow in mountainous districts from this circumstance, is familiar to every one. If the mountains are very lofty they retain their capping of snow throughout the year.

(352.) The appearances presented by fresh-fallen snow, when the fall has been considerable, are extremely beautiful, and calculated to arrest the attention of all admirers of Nature. If the wind blow, it often sweeps the snow on the hedge-banks into the most graceful curls and arches, or ripples it into a succession of undulations; if the air be still while the snow is falling,

* *Climate of London*, vol. ii. p. 225.

the latter then gathers upon the boughs and twigs of trees, clothing them with a crystalline foliage of the purest white. But this accumulation of snow, during frost especially, is very hurtful to tender evergreens; and the owners of ornamental grounds and gardens, who value their shrubs, will do well to cause the snow to be lightly shaken off them by a fork, as soon as possible after it is down. Not only are the branches bent down, and often broken, by the weight of the snow, but the whole shrub is liable to be killed or greatly injured by the alternations of temperature to which it is afterwards exposed, while the snow is still clinging to the leaves. If the sun shine, and the shrub is so situated as to receive its heat, the snow, in melting, absorbs caloric from the plant : after the sun's influence is withdrawn, the partially melted snow is converted into ice, having an equally injurious effect upon the leaves. It is this "repeated melting and freezing of the snow that," as White observes in his *Natural History of Selborne*, "is so fatal to vegetation, rather than the severity of the cold*." Shrubs, if they ultimately recover from such treatment, yet show their sickly condition for months afterwards. The leaves look as if they had been scorched. Of course, too, those shrubs which are planted in the warmest aspects suffer the most in this way. This was remarkably the case in my garden at Swaffham Bulbeck after the severe winter of 1829–30, the shrubs that sustained the greatest injury being Bays, Laurustines, and Laurels. And in one or two instances, in which these last grew on the north side of a large round clump, but had overtopped the other shrubs, it was particularly observable that

* Letter LXI. to D. Barrington.

the uppermost branches, which from this circumstance were occasionally exposed to the full action of the sun, turned of a dead yellow, whilst the rest of the plant remained green.

(353.) Howard has observed that "when the sun shines clear, and the temperature is at the same time too low for it to produce any moisture, the level surface of snow may be found sprinkled with small polished *plates of ice*, which refract the light in colours as varied and as brilliant as those of the drops of dew. At such times, there are also to be found on the borders of frozen pools, and on small bodies which happen to be fixed in the ice and project from the surface, groups of feathery crystals, of considerable size, and of an extremely curious and delicate structure." He adds, that "from the moment almost that snow alights on the ground, it begins to undergo certain changes, which commonly end in a more solid crystallization than that which it had originally. A notable proportion evaporates again, and this at temperatures far below the freezing-point." Howard found, on exposing one night "a thousand grains of light snow, spread on a dish (which had previously the temperature of the air) of about 6 inches diameter," that "in the course of the night the loss was about sixty grains*." This evaporation of a portion of the snow, which has been before alluded to (200), would probably, of itself alone, have a tendency to consolidate the rest, by lowering its temperature while in the act of passing into the vaporous state, without of necessity the temperature of the air being lower than when the snow descended. Every one knows how hard snow becomes as the cold increases, and within how

* *Climate of London*, vol. ii. p. 291.

short a period after it is down it feels, to a certain degree, compact under the feet, compared with its soft and yielding condition if walked upon while it is yet falling.

(354.) There is a circumstance connected with the melting of snow, which must have fallen under the observation of many persons, though it has seldom perhaps attracted such attention as to lead to the consideration of its cause. I allude to the fact of its first melting about the bottoms of stems of trees, or around plants and low bushes, so as to leave these bodies, after a time, standing in the midst of more or less deep excavations, while the snow elsewhere is still undiminished in depth. This has nothing to do with any heat, above the temperature of the air, which vegetables may be supposed to possess in the living state; for the effect is the same if we simply fix a stake or a common walking-stick into the ground. In this case also, after the snow has ceased falling, and a certain interval has elapsed, more especially if the sun shine brightly, the snow will no longer be found in close contact with the bottom of the stake. Yet it is not owing to the *direct* rays of the sun that this takes place, or there would be no reason why the snow should not waste by melting, everywhere the same, under the sun's influence. The real explanation of the phenomenon is that the solar heat is first communicated to the stake, or other bodies above mentioned, and *from them* radiated to the surrounding snow; the effect being greater in this case, in consequence of more calorific rays being absorbed by the snow from these secondary sources, than when it receives them directly from the sun itself*.

* The above explanation of this phenomenon, which I had

(355.) In like manner, we may observe that the snow first melts beneath the branches of trees which project out horizontally to a considerable distance from the stem. In such situations, the ground, within a limited space, will be cleared sooner than in more open spots. Here, however, there are two causes combining to bring this about. Not only does the snow, where thus sheltered, receive the influence of the sun's heat radiated from the boughs and other parts of the tree, but it is protected at night by these same boughs from *terrestrial* radiation, which lowers the temperature of the snow in other places, rendering it less ready to be dissipated under the influence of any elevation of the temperature of the air.

long ago noticed, without rightly understanding the cause, is given from M. Melloni, whose memoir "On the cause which produces the speedy Melting of Snow around Plants," will be found in the *Edinb. New Phil. Journ.* vol. xxv. p. 242. He details several experiments made by himself in connexion with this subject, tending to show that his explanation is the right one.

CHAPTER V.

THUNDER-STORMS.

(356.) It is not my intention to enter upon the general subject of atmospheric electricity, which is as yet but imperfectly understood, and in reference to which I have made no particular observations. I shall here speak simply of some of its effects as shown in ordinary thunder-storms. Of the identity of lightning with the electric fluid no doubt is entertained at the present day, or that thunder-storms depend upon the electrical condition of the atmosphere in relation to the earth and clouds. It is thought that the chief source of atmospheric electricity is evaporation, though there are many other sources exerting a less considerable influence in the production of it. Though "the conversion of pure water into vapour, at any temperature, is not attended with any disturbance of the electric equilibrium," it is otherwise with water containing any alkaline, acid, or saline substances in solution. In this case the vapour rises with signs of positive or negative electricity, according to the nature of the substance dissolved in the water. The surface of the sea, therefore, in the course of evaporation, and not the sea only, but all other natural collections of water, such as lakes, rivers, &c., none of which are absolutely pure, must yield copious

supplies of atmospheric electricity.   The same must be
the case with the moisture evaporated from the humid
soil*.   Clouds being collections of condensed vapour,
will exhibit more or less of this electricity, the quantity
varying with the climate, the season, and the time of
day.   They are sometimes positively, and sometimes
negatively electrified.   If their electrical conditions are
similar they will repel, if different they will attract each
other; and if in this last case they are separated by a
stratum of dry air, " the equilibrium will be restored,
when they are so near each other that the force, or
intensity, of the electricity overcomes the resistance of
the intervening air, by a violent discharge, producing
the phenomena of thunder and lightning †."   In other
cases the discharge takes place in a downward direction,
in consequence of the opposite electrical states of a
thunder-cloud and the earth ; or more rarely, as has been
confidently stated by some, in an ascending one from
the earth to a cloud.   Owing, however, to clouds being
imperfect conductors of electricity, the equalization is
seldom effected at one effort, a succession of discharges
being necessary for the purpose.

(357.) *Lightning.*—The different appearances which
lightning assumes seem greatly to depend upon the
arrangement of the clouds, and the direction the lightning takes.   Much of the lightning that is seen during
storms, even when the storms are very heavy, only
passes from one cloud to another.   If these clouds are
not of the same height, they are irregularly illumined,

---

* See an article "On the Electricity of the Atmosphere," in
the *Companion to the Almanac* for 1839, from which the above
remarks are borrowed.

† Drew (*Practical Meteorology*), p. 218.

though often in the most brilliant manner, without our
being able clearly to discern the flash, or to follow it in
its path. Sometimes the whole heavens are illuminated
at once in this way, the effect of which is very striking
at night, the landscape around, for half a second or
more, standing out in bold relief, almost as distinct
as day, and then as suddenly falling back again to
pitchy darkness.

(358.) When the lightning is seen direct, without
the intervention of clouds, as when it passes from one
cloud to another, both clouds being at the same elevation, or when it takes a downward course to the earth,
it then appears as a luminous streak of dazzling brightness, the light being either white, or inclining to violet.
Often, when descending to the earth, it is compound or
branched, as well as acutely angled, in which case it is
called *forked* lightning. The zigzag course taken by
such lightning is probably due to slight differences in
the constitution of the different strata of air through
which it passes, rendering their conductibility unequal.

(359.) *Sheet*, or *silent* lightning, as it is sometimes
called, from its being unaccompanied by thunder, is
that which occurs in warm summer evenings, always
near the horizon, and is generally thought to be the
reflexion of a storm *below* the horizon. Sometimes the
flashes are very diffuse, spreading upwards, as well as
laterally, to a wide extent. They have a peculiar flick-
ering appearance, different from what is observed when
the lightning itself is seen direct; though in fact this
is more or less the character of all lightning rendered
indistinct by the intervention of clouds, even in storms
close by the spectator.

(360.) Sheet lightning is generally reflected upon

clouds situated in the direction from whence it comes.
It is often, however, seen when no clouds are visible
above the horizon; in which case it must either be
reflected on the atmosphere, in the same manner as
twilight is, or it must proceed from some other source
than a storm below the horizon. Accordingly it has
been questioned by some, whether lightning without
thunder does not sometimes arise from a peculiar electric state of the air, independent of storms altogether.
I remember one instance myself, on the last day of
April, 1826, in which, happening to be out between 10
and 11 at night, the sky at the time being perfectly
serene and cloudless, and the stars shining bright, I
saw three distinct flashes of lightning at short intervals, in the direction of S.E., and at a considerable elevation above the horizon. They were similar to flashes
seen direct in an ordinary storm, not presenting that
diffused and indistinct light above described. No
thunder followed. The air was very cold: the exact
temperature was not noticed, but the thermometer had
been only 47° a few hours previous. The barometer
was above 30 inches, and the wind N. I could not
learn afterwards that any storm had occurred elsewhere, to lead to the belief that the flashes I saw were
only the reflexions of lightning below the horizon.
Indeed if they had been, they would hardly have appeared at the elevation at which I saw them, without
the light extending *down* to the horizon*.

* Humboldt speaks of lightning without thunder in South
America, though in the instance mentioned by him the sky was
in part much clouded. He says, " We left the Island of Panu-
mana at 4 in the morning, two hours before sunrise. The
sky was in great part obscured, and lightnings furrowed thick

(361.) There can be no doubt, nevertheless, that the summer lightning seen near the horizon, generally speaking, is caused by a distant storm, too far off for the thunder to be heard. And the distance at which the reflexion of lightning is visible is very great. Dr. Hooker, when in Tibet, saw the sky illumined with flash lightning, which he believed to be the reflexion of a storm "which raged in the plains of India, beyond the Terai, fully 120 miles, and perhaps 150, south" of his position*. Sir J. Herschel says sheet lightning may be seen at the distance of 150 or 200 miles.

(362.) Besides the above different kinds of lightning, it sometimes appears in the form of balls or globes of fire, which are remarkable "for the slowness and uncertainty of their motions," as well as for their destructive effects when they burst. The nature of these *fireballs*, as they are called, is very little understood†. I never saw an instance of one myself.

clouds at more than 40 degrees of elevation. We were surprised at not hearing the sound of thunder; was it on account of the prodigious height of the storm? It appears to us, that in Europe the electric flashes without thunder, vaguely called heat lightning, are seen generally nearer the horizon."—*Travels*, vol. v. p. 8.

See also a notice of vivid forked lightning, without thunder, in *L'Institut*, 1840, p. 100.

* *Himalayan Journals*, vol. ii. p. 161.

† Arago considered them as "among the most inexplicable phenomena in physics." He says, "These balls, or globes of fire, appear to be agglomerations of ponderable substances strongly impregnated with the very essence of the thunderbolt. How, then, are these conglomerations formed?—in what regions are they produced? Whence do they obtain the substances which compose them? What is their nature? and why are they sometimes suspended for a long period, only that they may

(363.) *Thunder.*—Thunder has been explained by Kämtz as "the noise resulting from the displacement of the air by the electric spark, and irruption of the surrounding air, which fills up the vacuum formed, as it happens when we open a case that closes well." But Arago regards this explanation as unsatisfactory, stating that no answer has hitherto been returned to the inquiry, "by what physical cause is it that the lightning produces a vacuum?" He considers that the right explanation of thunder has still to be discovered.

(364.) The length of the interval that elapses between seeing the lightning and hearing the thunder, depends upon the distance at which the cloud, whence the lightning proceeds, is from the spectator*. And, other

precipitate themselves with the greater rapidity? &c. &c. To all these inquiries, science remains mute, and can make no reply."—*Edinb. New Phil. Journ.* vol. xxvi. p. 85. The following circumstance, mentioned by Howard, is an instance of the *slowness* with which these fire-balls move in some parts of their course. Speaking of lightning, which, during a storm, "struck a chimney, which it threw down, bringing a quantity of brick and sand into a room, in which a party of eight persons were at tea,"—he adds, "I was informed by one present, that *after the first explosion* a ball of fire made its appearance under the door opposite the chimney, *where it remained long enough for the whole company to notice it.* It then moved into the middle of the room, and separated into parts, which again exploded like stars from a rocket."—*Climate of London,* vol. ii. p. 85.

* "According to the most recent experiments, the temperature of the air being 50° Fahr., the velocity of the sound equals 368·5 yards (English) per second. If the cloud, whence the lightning has issued, is at a distance in a straight line of 368·5 yards, there will elapse *an entire second* between the appearance of the light and the arrival of the sound. Hence, then, an observer who shall have determined with his watch, the number

circumstances being the same, the clap is louder as
that distance lessens. But there are some very observable differences in thunder, arising in part from
other causes. Sometimes it breaks upon the ear as
a rolling noise, getting first louder each moment, then
weaker, until it ceases to be heard; at other times we
hear only one loud report;—or it begins with a loud report, followed by a rolling peal that dies away gradually;—or there is a succession of loud reports, the
rolling, instead of diminishing, " seeming, on the contrary, to gather force from time to time, and appearing
intermixed with more violent claps, like the noise produced by a mass of something falling down stairs*."
Two circumstances probably combine to bring about
these differences. In many instances, the rolling peals,
which are so prolonged after the first crash of thunder
has taken place, are due to the clouds reflecting the
sound, and causing a succession of echoes, similar to
what we hear when a gun is fired in a hilly district.
The length of time for which the thunder is heard, will
then, so far as this cause is considered, depend upon the
number and the arrangement of the clouds. But this
is not sufficient to explain all the variations in the intensity of the sound. To understand these, it must be
remembered that thunder proceeds successively from

of seconds comprised between the arrival of the flash and that
of the thunder, may easily deduce the distance at which he is
from the point from which it emanated. All that is required
is, that he shall multiply this number, whether a whole or a
fractional one, by 368·5 yards: the product will be the distance
sought after, expressed in yards."—*Arago, l. c.* p. 92. On the
above calculation the sound of thunder traverses a mile in rather
more than four seconds and three quarters.   * Kämtz.

every point in the path along which the lightning travels. And this path differs in length and direction at different times. If the lightning be of a simple kind, and "confined to a mere point in space," a clap of thunder would ensue, which, however loud it might be, would be of no continuance. If, on the contrary, the flash be a compound one, extending to a considerable distance, the noise of the thunder will be prolonged in proportion to that distance. If, in this latter case, the direction which the lightning takes be a nearly straight and horizontal one, as when the latter traverses clouds at the same height, and be wholly away from the observer, the roll of the thunder will be every moment getting less, until it ceases altogether. But if its path be zigzag, as in the case of forked lightning passing from a cloud to the ground, some points in its course will be nearer to the spectator than others, causing the thunder to be louder at intervals, instead of dying away gradually. Should the lightning take a circular path around the observer, as the arrangement of the clouds will sometimes favour its doing, "the noise will then reach his ear from every point at once, and a stunning crash be the effect."

(365.) It is a singular circumstance that thunder, notwithstanding its loudness, should be heard to such a much less distance than the report of cannon. According to Arago, "the greatest distance at which thunder has ever been heard, is little more than fifteen miles, and the greatest ordinary distances do not amount to more than about ten miles:" whereas, he mentions an instance in which the report of cannon has been heard upwards of ninety English miles. He is disposed to attribute the more rapid dying away of the sound of thunder to "the partial reflexions to which it is subjected in obliquely encountering the different substances which separate the atmospheric strata of different densities."

(366.) As lightning sometimes occurs without thunder, so we sometimes hear even loud thunder without any lightning being seen previously. In many instances in which this is the case, there are probably two strata of clouds, one above the other, the uppermost being charged with the electric fluid. Flashes are then transmitted to the lower cloud, which, if its density be considerable, does not allow their light to pass through it, in sufficient quantity to reach the eye of an observer placed beneath. But the same cloud offers no obstacle to the passage of sound, and the thunder falls upon his ear as loudly as at any other time *.

* In drawing up the above remarks on thunder and lightning, some assistance has been derived from Arago's Essay on the subject, published in the *Annuaire du Bureau des Longitudes*, for 1838, a translation of which will be found in the *Edinb. New Phil. Journ.* vol. xxvi. pp. 81 and 275. To this Essay the reader is referred for many interesting as well as singular details connected with these phenomena. There are some curious facts mentioned, in particular, relating to the effects of lightning on different bodies struck by it. Thus, in the case of persons exposed to lightning, he says "numerous instances may be cited, in which it would seem, that some individuals appear to have been preserved, and others struck, according as they wore particular garments, manufactured of particular stuffs." He considers it as an ascertained fact, "that wax-cloths, and silk and woollen stuffs are less permeable to the material of lightning than linen, hempen cloths, or other vegetable substances." It would also appear "that animals may be more or less severely injured in different parts of the body, according to the *colour of the hair* which covers them. Thus, at the beginning of Sep-

(367.) *Circumstances under which thunder-storms oc-cur.*—There are certain conditions of the atmosphere,

tember, 1774, an ox was struck by lightning at Swanborrow, in Sussex. The colour of this animal was red, spotted with white. After it was struck, all were surprised to observe *the denudation of the white spots*; on these not a single hair remained, whilst the red portions of the hide had not undergone any apparent alteration. The owner of the animal stated, moreover, that two years previously, another ox, under the same circumstances, had exhibited precisely the same appearances." The case is also mentioned of "a pie-bald horse having been struck by lightning, at Glynd," and its owner remarked afterwards, "that, throughout the whole extent of the white spots, the hair came off, as it were, of itself, whilst, in the other parts, the coat adhered as usual." "Men" are said "to resist the effects of lightning more power-fully than horses and dogs." There is also a great difference in men in this respect, depending in part, as Arago thinks, "upon the physical constitution of the individual." There are some persons not conscious of an ordinary electric shock, transmitted through them from the conductor of an electrical machine. Such individuals—exceptions to the general rule—he considers "must be ranged among *non-conducting* bodies which the light-ning respects, or which it, at least, strikes less frequently."

Arago also proposes the singular question, whether "the light-ning strikes before it becomes visible?" He thinks it possible that the electric fluid itself may traverse space with even a greater velocity than that of light. He grounds the supposition on the circumstance of many persons, who had been struck down, and for a time rendered unconscious, by lightning, having, on their recovery, been ignorant of the cause; and de-clared that, previous to being struck, they had *neither seen the lightning nor heard the thunder.* If this fact be established, it would follow "that nothing is to be apprehended when the flash has been seen."

The power which lightning possesses of splitting heavy masses of stone and other matter, and projecting the fragments to a distance in all directions, is probably due to the development, which, if not absolutely necessary for the production of a storm, are almost always present, before it comes on, and at least warn us that a storm is not improbable. This remark, however, applies only to the storms of summer. The conditions are, great heat, stillness of the air, a serene sky, and much humidity in the soil and lower strata of the atmosphere. A high dew-point, especially, is always to be suspected in hot weather, when combined with a dead calm, and the wind in the W. or S.W., notwithstanding a fine morning. If there has been much rain on the previous day, and ill-defined large dark masses of cloud begin to show themselves about noon, the result may be yet more certainly anticipated. The barometer often stands as high as 30 inches, or not much under, previous to a thunder-storm, and therefore affords no trustworthy indication one way or the other.

(368.) It is the above state of the atmosphere that induces the peculiar feeling of lassitude, so often experienced before a thunder-storm. The air is oppressively hot, while its humidity, approaching to saturation, and the absence of all wind, prevent any evaporation from the surface of our bodies. It is not uncommon for persons to consider this languor as due to the electricity by which a storm is accompanied, or, as we often hear " in the very substance of those bodies which it traverses, of some eminently elastic fluid," perhaps "steam." It is owing to this action of aqueous vapour, being the sudden evaporation of the sap, that wood is so singularly torn into shreds, when trees are struck by lightning.

The action of lightning on gunpowder seems uncertain. Though it may generally set fire to it, Arago mentions two instances, in which lightning fell upon powder-magazines, in the first case shattering the casks, in the second overturning them, *without igniting the powder.*

it expressed, to " thunder in the air." Those, especially, who are not strong, or in delicate health, are apt at such times to suffer from headache, nausea, or other slight derangements of the system, which they at once attribute to the above cause. But this is probably a mistake. It is not meant that the human health may not be affected in some cases by a powerful development of atmospheric electricity,—though medical science is hardly perhaps in a condition to say how or to what extent,—but all the above symptoms may be easily explained by that relaxation of the system, which heat and checked perspiration naturally produce, without having recourse to any such agency.

(369.) Neither is the development of electricity in the air a circumstance attendant on thunder-storms alone. More or less is evolved on all occasions of condensation of vapour, whether the condensation take the form of fog, hard showers of rain, hail or snow. It is probable that there are many states of the atmosphere, in which the intensity of the electricity diffused through it is not far short of what occurs during thunder-storms, though there may be no actual thunder and lightning to manifest its presence in that marked way. Kämtz, too, will tell us, that, looking to " all the circumstances that accompany the development of electricity, we must consider the condensation of vapours as the cause of its production, and conclude that it is the storm that produces the electricity, and not the electric tension that produces the storm, as is the general opinion*." This at once shows how erroneous it must be to refer any of the above feelings to a cause, not yet in operation, if the thunder-storm has not yet commenced.

* *Meteorology* (by Walker), p. 364.

(370.) In like manner, it is probably a mistake to suppose, though the idea is so commonly entertained, that thunder has the effect of tainting meat, and turning beer or milk sour, &c. These changes in articles of food, however rapidly they may be brought about, may all be traced to that same state of the atmosphere, which exercises its influence in the way above alluded to on the human subject. It is the high temperature combined with great moisture that favours the putrefactive process, and causes liquors to ferment,—and these effects would equally follow, under such conditions of the air, whether attended by a thunder-storm or not.

(371.) *Seasons of occurrence of thunder-storms.*— Though thunder-storms are most frequent in summer, they are by no means rare in the early part of the spring, especially in March, when I have several times known them to occur during the rough weather that is so characteristic of that month. Kämtz observes that there are two classes of storms, "the one class being due to the action of an ascending current, the other the result of the conflict of two opposite winds*." While the former class comprises the storms that prevail in summer, there can be no doubt that to the latter belong the storms at the period of the year I am now speaking of. These storms generally come from the N.W., the wind often blowing strong at the time, accompanied by hard showers of hail and rain. The temperature has been low, seldom much above 40°, though the flashes of lightning have been as vivid, and the peals of thunder as loud, as at other seasons†.

* *Meteorology* (by Walker), p. 362.
† It has been observed that plants come forth and bud with increased rapidity during spring, after thunder-storms; which,

(372.) In one instance, I even remember a heavy thunder-storm occurring in the middle of winter. This was in January 1841. The weather had been very severe, with continued frost, all the latter half of the previous December. A thaw took place at the beginning of the new year, and on the morning of the 3rd of January, between five and six o'clock, this storm came rather suddenly up from the S.W., with torrents of hail and rain, the wind blowing a gale at the time, and the temperature rising to 43°. The barometer fell during the night more than seven-tenths of an inch. The storm lasted about half an hour. Either during the time of its raging, or shortly after it was over, the wind shifted to the N., and the temperature again fell; and in the afternoon of the same day there was a recurrence of snow-storms and frost. A few days later in the month, the frost became more intense than it had been previous to the storm, the thermometer, on both

---

if it be a well-established fact, would serve to show the powerful influence which atmospheric electricity exerts upon the vegetable world.—*Agassiz.* Arago mentions one very remarkable instance in which vegetable life would seem to have been promoted by the effects of lightning. He says,—" There existed a few years ago, between Tours and Rochemort, a château, called Comacre, with an avenue of fifteen hundred poplars. One of these was struck by lightning, which left evident marks of its action, both on the trunk of the tree and on the ground adjacent to it. From thenceforth the growth of this tree became something quite peculiar : the dimensions of its trunk soon surpassed those of any other tree in the avenue, to such a degree that the difference attracted the notice of the least attentive, and of those who were entirely ignorant of the event which had occasioned it." — *Meteorological Essays* (Sabine's Translation), p. 258.

the 8th and 9th, falling as low as 6°, nearly the lowest temperature I ever registered.

(373.) As thunder-storms are most frequent in summer, so also they generally occur about, or soon after, the hottest time of the day. It is then that the quantity of moisture evaporated from the ground, and taken up by the ascending current into the region of the clouds, attains its maximum, while there is a corresponding accumulation of electricity arising from its condensation. Sometimes, indeed, the moisture may be again evaporated, if the air in the neighbourhood of the clouds be in a sufficiently dry state to take it up; otherwise the mass of condensed vapours keeps increasing until a portion of them is precipitated. Whether the rain take the form simply of showers,—in which case the drops serve to reconduct the electricity to the earth,—or that of hard storms, accompanied by thunder and lightning, will depend upon the degree to which the electricity is developed. If its intensity be very great, the electric fluid becomes, as it were, impatient of restraint, and, without waiting for any ordinary passage of escape, opens for itself at intervals a communication with the earth or some other cloud, at the same time accelerating the rain, and causing it to fall more heavily than before.

(374.) From the above circumstances there results a remarkable periodicity in storms. In certain states of weather, they recur regularly at the same hour for a number of days in succession. The mornings are fine, with a cloudless sky and a serene atmosphere; but after the sun has attained a certain altitude, evaporation from the soil, always active during the hot season, goes on briskly; cumulus clouds soon begin to form,—these after a while pass into the heavier and darker forms of cumulostratus,—which latter gradually spread and grow in all directions, until about three or four in the afternoon they resolve themselves into rain. After one or more of these storms, the air speedily clears; as the temperature declines, the vapours again sink downwards to the earth, and a bright cloudless night ensues, to be followed next day by a repetition of the same phenomena.

(375.) Such weather will continue until there is some decided change of the wind below, or shifting of the currents in the higher regions of the atmosphere, to bring about an alteration. So long as it lasts, the barometer will generally be found considerably below the mean, showing little disposition to rise. For though, as before stated, it may often be high before the occurrence of a particular storm (367), this is hardly ever the case when the weather continues changeable for some days.

(376.) Though the circumstances under which thunder-storms occur are usually connected in the manner above described, with the hottest period of the day, the atmospheric conditions which sometimes cause the morning of a summer's day to be as sultry as the afternoon, will also similarly bring thunder-storms about at a very early hour. Thus, on the morning of the 3rd of July, 1845, during very changeable weather, the temperature rose $10°$ higher than it had been any part of the preceding day, the dew-point being also very high; and at nine A.M., a heavy thunder-storm came up from the S.W., which lasted nearly four hours. After it was over, the thermometer stood at $80°$, and had been very little less all the time the storm was raging.

(377.) *Awful thunder-storm.*—Heavy thunder-storms

are not unfrequent at Swaffham Bulbeck, but I have observed that they very seldom pass directly over the village. Either they come up from the W. or S.W., and pass off to the N.W. over the fens in the direction of Ely; or else they come from a point nearer the S., and divide into two before reaching the village, one portion taking the direction just named, and the other an easterly one towards Newmarket Heath and the Woodlands. One most terrific tempest, however, which occurred in Cambridge, and extended over a wide tract of country for miles round, on the 9th of August, 1843, deserves to be especially mentioned. I have already more than once alluded to this thunder-storm, which, for its violence and destructive effects, far exceeded any known before in that part of England. Unfortunately I was not a spectator of it. I had left home the day previous, and was at Peterborough at the time. Even there the storm was exceedingly heavy, though not of such an unusual character as to call for any particular details. Far different, however, was it in the district from which I had come. As it is part of the object of this work to give a record of all such phenomena, so far as connected with the neighbourhood in which I was then resident, and having no observations relating to the storm in question of my own to offer, I shall make no apology for copying from the ' Cambridge Chronicle,' published the Saturday after the day on which the storm happened, the following account of the principal circum-stances by which it was attended, along with a general statement of the damage it occasioned. It may, too, be desirable that this account of a storm of such unparal-leled magnitude should not be confined to a mere news-paper.

" To-day it is our painful duty to record a storm of thunder and hail more terrific in its character, and more disastrous in its results, than any by which this district has been visited within the memory of living man, or indeed of which history supplies us with an account. Wednesday, the 9th of August, 1843, will hold a conspicuous place in the annals of this and the adjoining counties, and the remembrance of it will never recur without sorrow to the minds of those now living, for to many it brought positive ruin, while very few escaped more or less of the injury with which it was fraught. Experience totally fails to supply us with anything in this latitude approaching the devastation occasioned by the tempest of which we speak, or the terror which it diffused amongst all classes of people during the time of its continuance.

" The early part of the day was hot and close, but there was nothing to indicate the approach of such a storm as afterwards visited us. About two o'clock there were symptoms of a change in the sunny atmosphere which we had in the morning: clouds began to gather, and the air felt heavy and oppressive, and between two and three o'clock, the rumbling of distant thunder was heard. This increased, and was accompanied by-and-bye with very vivid flashes of lightning, but for a long time no rain fell. In the meantime the atmosphere became darker and darker, and it was evident that a storm of extraordinary magnitude was at hand. Large drops of rain began to fall soon after four o'clock, and in a short time a perfect deluge poured down upon the earth. The hail-storm began at about a quarter before five, at which time there was a tolerably brisk wind from the N.E. Whether we regard the size of the hailstones,

the violence with which they were driven against objects on the earth, or the destruction they have caused,
there is no parallel in this part of the world to the tempest which now raged for nearly half an hour. The
extraordinary darkness of the atmosphere, with the
clouds almost sweeping the house-tops, the incessant
roar of the thunder and flash of the lightning, and the
deafening noise of the falling hail, impressed one with
a sense of awe and admiration which cannot be described.

" The scene was positively terrific, and the fright of
many of the inhabitants of the town was in no small
degree increased by the crash of broken windows, and
the inundation of their houses. During the whole of
this time it was impossible for the eye to penetrate
many yards through the storm : the hail fell with such
wonderful closeness, and there was such a peculiar mistiness rising from the earth, that a complete barrier was
opposed to the power of vision. We are almost afraid
to speak of the size of the hailstones, or rather blocks
of ice, but we are certainly not exaggerating in the least
degree, when we say that very many of them were as
large as ordinary walnuts : some indeed far exceeded
this size; one that was picked up measured three and
a half inches in circumference, and several have been
described to us as being about as big as a pullet's egg.

" Mr. Glaisher, of the Cambridge Observatory, has
been so obliging as to furnish us with the following
observations on this storm, made at that Institution :—

" ' This day will be for a long time memorable, on
account of the extraordinary storm and accompanying
phenomena which occurred in the afternoon. The day
previous was sultry, and so was this, but hardly suffi-

ciently so to make the approach of any storm of consequence expected.   The morning of this day was fine,
with sunshine; the wind until 11 A.M. was from South,
after then from North.   At 2 P.M. thunder was indistinctly heard in the direction of N.W., and from this
time until four o'clock, the distant, though gradually
approaching storm, gave unequivocal signs of being of
more than ordinary magnitude.   The lightning, or
rather its reflexion, was first seen in the W. and N.W.
horizon about 3 o'clock: by $3^h 30^m$ the lightning itself
was visible, and the thunder was then heard in more
distinct claps, the previous sounds being but the echoes
from the masses of clouds which hovered about; these
echoes were so numerous that the reverberations of one
clap had not subsided before the next occurred, thus
causing a continued rumbling since 2 o'clock.

" ' From 4 o'clock until $4^h 45^m$, the storm approached
rapidly in an almost due east direction, passing therefore rather northward; some large drops of rain fell in
this interval, and the flashes of lightning became very
vivid and of a brilliant purple colour.   At $4^h 45^m$, the
hail-storm began, and for twenty minutes continued
with a violence probably unprecedented in the last century, in the latitude of England.   The great size of the
hailstones was the first thing to attract attention, for
many measured *an inch in diameter*; some were even
larger, and the average size was probably from half to
three-quarters of an inch in diameter.   They fell as
closely as the drops of rain from a waterspout, and this,
with their weight, and some accelerating force from a
brisk N.E. wind, caused them to do immense destruction.   The temperature of the rain was certainly not
higher than 40 degrees; the hail was, of course, icy;

it fell upon the earth whose temperature was considerably higher, and thus a mist, or almost a steam arose, and made the view still more dreary than the effect of a snow-white ground alone would have presented.

" ' The entire storm went by N.E. to E., and for a time disappeared, though the lightning now and then became visible: at 6 o'clock it had completed three-quarters of a circle, and appeared again in the S. and S.W.; soon afterwards it was evident that Cambridge would have its near approach again, although it was fast getting Westward. The character of the lightning in this second appearance was more terrific than before, for the principal portion of every flash was in a vertical direction, and on many occasions several of these vertical streams were visible almost simultaneously; once I counted seven distinctly, at irregular intervals, varying from 4 to 10 degrees : from this circumstance it is much to be feared that considerable destruction has occurred from this cause also. No hail accompanied this appearance of the storm, but a very copious rain fell whilst it was passing from W. to N. Frequent flashes of lightning were visible all the evening, in the N. and N.E. horizon, and there were frequent heavy showers during the night, especially about 1 o'clock, when thunder was again heard distinctly. Unfortunately my rain-gauge was not in a proper condition to receive so much rain without overflowing, and therefore I cannot state accurately the quantity of rain which fell; but estimating from so much as the gauge did retain, and some years' experience, I think the amount cannot be less than $2\frac{1}{2}$ inches. The wind varied but little from the N. except after the hail-storm, when it blew

from the E. for an hour, and then returned to N.   The
barometer fell but little before it commenced, and then
continued stationary.'

" When the first storm abated, which it did about a
quarter past five, the aspect presented on looking round,
was dreary and distressing to the highest extent.   The
streets and roads were like running rivers ; Midsummer
Common was one sheet of water, in the midst of which
a burst drain boiled up like a miniature Icelandic gey-
ser ; glass wherever exposed to the fury of the elements
was of course shivered to atoms, and with such extreme
violence had the hail descended, that in some instances
it passed through windows almost like a ball from a
pistol; trees were half-stripped of their leaves, which
were seen lying in layers on the roads, and on apple-
trees the fruit was battered to pieces, in some cases
pieces being actually scooped out; birds, even rooks
and pigeons, were killed in large numbers, and picked
up in the country in all directions : the houses in the
town were in many parts flooded, the cellars being
several feet deep in water, which had also made its way
through the roofs, and destroyed furniture and stock-
in-trade in its course: garden-produce was of course
utterly destroyed, and the havoc made amongst the
crops just ripening for the sickle was of the most la-
mentable kind.   Herein, indeed, the devastation effected
by this awful storm is greatest and most to be regretted:
harvest had just begun ; the bounties of the earth were
waiting to be gathered and stored for the comfort and
sustenance of man, and in one short half-hour they were
swept away or rendered perfectly useless.   The effect
on the crops where the storm was most violent was very
remarkable.   In some instances the straw was actually

beaten down and broken up into little pieces, almost as if it had been *chopped*, and the ears were as bare as if they had been regularly thrashed *."

(378.) The damage done by this storm in the way of breakage of glass, and destruction of garden and field crops, was almost incalculable. It was supposed that in the University and town alone it might be set, at the very lowest, at £25,000. What it amounted to in the surrounding parishes it was quite impossible to say. The storm seemed to have raged with most violence at Quy, in which parish one farmer alone was stated in the papers to have suffered to the extent of £2000. At Bottisham and Swaffham Bulbeck, the storm was rather less violent, and the damage less; but even here, when I returned home a few days afterwards, in many places, the corn-fields, which had standing crops in them at the time of the storm, looked as bare and beaten as they ordinarily do after the corn and stubble have both been cut and cleared away.

(379.) It will give some idea of the immense quantity of hailstones that fell during this storm, when I state that a friend of mine, who had gone into Cambridge from Swaffham Prior, in a four-wheel pony carriage, that morning, and who was kept there while the storm was raging,—on his return home in the evening, three hours after the storm had abated, found the hailstones, lying still unmelted, in such heaps in the road between Quy and Anglesea Abbey, that his horse was unable to drag the carriage along. On getting out to make a track for the wheels, he told me *he sank up to his knees.* Representations of two different forms of these hail-stones, of the natural size,—some of which were so

* *Cambridge Chronicle*, Saturday, Aug. 12, 1843.

large, that they stuck in a wine-glass—are given in the
annexed woodcut.

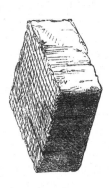

(380.) The above storm was not confined to Cambridgeshire, nor even to the adjoining counties. Before
reaching that neighbourhood, it had travelled over a
large part of England, from W. to E., dividing in one
part of its course into several branches, which took different directions. Almost all the midland, as well as
some of the southern counties, were visited by it during
some part of the day; though in very few instances its
violence in other places was equal to what it was in
Cambridge*.

* See Lowe's *Atmospheric Phenomena*, p. 201, where there is
a statement of the exact course this storm took, as well as many
details respecting it, collected from different observers at the
principal places it visited.

CHAPTER VI.

GENERAL OBSERVATIONS ON THE WEATHER.

MANY remarks bearing on the subject of this chapter will be found dispersed through the preceding parts of this work. Nevertheless it may be useful to bring them in some measure together, under a distinct head, previous to giving a few directions for judging of the weather beforehand, so far as it is possible to form a judgment at all.

(381.) It might seem at first almost hopeless to attempt to describe the different kinds of weather, so insensibly do they pass one into another. Between the extremes of hot and cold, dry and wet, serene and cloudy, there is every conceivable shade of variation. Nor do these several conditions of atmosphere remain always relatively the same. Drought may consist equally with a high and low temperature : great humidity of the air may do the same. The sky, again, is often clouded for days together without a drop of rain falling; while it is never so serene as in the intervals of showers during changeable weather. Or a thick mist may rest upon the earth, drenching everything with wet, while but a few feet, comparatively speaking, above our heads, the sun is shining brightly, with every indication

of a fine day. The winds, too, are for ever shifting to
disturb an order of things, which, perhaps, but a few
hours before seemed settled and likely to continue.
From the alterations thus constantly taking place, in
this variable climate especially, in some one or more of
the circumstances of the weather, there is scarcely a
day, as it has been observed, which, perhaps, has not
"something different from all the rest in the year."
Still we may speak of the weather for meteorological
purposes, in the same way that we speak of it in reference to the purposes of common life. Every one knows
what is meant by fine settled weather, in which there
are no material fluctuations of temperature and pressure,
as shown by the thermometer and barometer, for many
days in succession, with a sky for the most part clear,
and of course no rain. By wet weather, again, we understand weather, in which the sky is more or less overcast, with steady rain at intervals, for a longer or a
shorter period;—by changeable weather, that which is
fine and wet by turns, in a fitful manner, the rain, when
it occurs, coming principally in the form of storms and
showers, which, perhaps, fall heavily while they last, but
which soon pass off, leaving the atmosphere as clear and
bright as before. In like manner, we speak of frost, as
distinguished from that mild open weather, which generally alternates with it in the winters of this country,
though sometimes the frost keeps its ground, without
any thaw to signify, for weeks together. Now it is not
attempted to describe the different sorts of weather,
otherwise than in reference to these general heads,
under which they may be arranged, further than is
requisite for pointing out the principal atmospheric
conditions usually attending them. We may then

easier mark any indications of a change about to take place.

(382.) The first thing to be noticed is that all changes of weather depend upon the winds. It is according as these blow, from what quarter, and whether they are steady or not, that we must draw our conclusions, primarily, about the weather. All our fine settled weather in this country is accompanied by steady winds; while, generally, in wet and unsettled weather the wind is variable, with two or more currents one above the other. It is further observable that the driest and most brilliant weather we ever experience, is the result of easterly winds, or of a wind blowing from some point in the N.–E. quarter. This has been before alluded to in the chapter upon Winds (144), where it was shown how much the seasons, as regards wet and drought, are influenced by the manner in which the Easterly winds are distributed through the year. Generally speaking, these winds blow steadily for a certain period during the spring, which is the time when they arrive at their maximum, and we have them only at intervals afterwards. But sometimes we have a much less share of them than ordinary at the former season, while they attain an unusual degree of frequency in the summer and autumn instead; or though they may occur in spring, instead of falling off at the end of May, they continue to a later period. Now it is an excessive prevalence of these winds, blowing steadily, or with little interruption, from the same quarter, that characterizes most of our dry summers, and which, from being dry, are of course also fine. In wet and changeable summers, on the other hand, the Easterly winds are mostly confined to the spring, or are at a minimum altogether

throughout the year. Likewise, during any period of
the year, it is the return of the wind to this quarter for
a longer or shorter interval, that gives rise to the most
settled weather, with a high barometer, and a sky often
bright and cloudless for days together from morning to
night. In the present case, however, it is supposed
that the whole body of the lower atmosphere, or to
above the height of that region within which the clouds
are ordinarily formed, is moving from the E. or N.E.,
and not merely a current next the earth, as indicated
by the vane.

(383.) But it is a more necessary condition for
settled weather that the wind be steady, than that it be
Easterly. We may have the weather fine and settled
for a time, with the wind in the N., N.W., S.W., or
we might add almost any quarter, if it does not vary.
From the S., however, the wind very rarely blows in
this country for any length of time together (140).
This and the S.W. are indeed the quarters from which
we generally have wet. But then the rain is due either
to the mixing of these winds with some other current
from a colder quarter, or to the warm vapours they
contain being precipitated by condensation on their
arriving at these latitudes. If they blow steadily, the
great body of the atmosphere moving in one and the
same direction, the first cause of rain is removed; and
if they continue blowing long enough, without any
material variation, though they may bring rain in the
first instance, after a time the redundant moisture will
be got rid of, and the weather clear. The air, too, in
some cases will have its temperature gradually raised by
the setting-in of a S.Westerly wind, and be thus enabled
to retain more moisture, without the latter being pre-

cipitated, than when that wind first began to blow. The weather therefore steadies itself by degrees, and continues fine, until some shifting of currents either in the higher or the lower regions of the atmosphere comes about to disturb the serenity of the sky, and to bring more rain or not according to circumstances.

(384.) During fine weather with a S.Westerly wind, the barometer will be steady like the weather, but it will not always be high. The wind in question is that, which, from being more rarefied than any other, is naturally attended by a lower state of the barometer. If the latter instrument be very high, we may infer that—though the current next the earth, and up perhaps to a considerable altitude, is from the S.W.,—there is still a large body of air coming from the N. or N.E. in the higher regions of the atmosphere, which gives the barometer its elevation, but which, from its flowing above the S.Westerly current below, without so far mixing with it as to lead to a precipitation, causes no interruption to the serenity of the weather.

(385.) If fine weather be accompanied by clouds, the latter are simply cumuli, which are never so definite and well characterized,—the most perfect being of a hemispherical form,—as when the weather is settled. The way in which these clouds are formed has been before stated (239); they are due entirely to evaporation from the ground by the sun's heat, rising in the morning, gradually increasing as the day advances, and subsiding in the evening. Their size, therefore, and indeed their very existence, is dependent upon the degree of moisture which the soil contains. When the wind is in the N.E., and especially after having been in that quarter for any length of time, its drying effect upon the ground is such as, to a great extent, soon to exhaust this supply of moisture. Or but little moisture is evaporated, beyond what the air is able to retain in the aëriform state, even to a considerable height, without being precipitated as clouds. The cumuli then get less and less each day, and after a time, if the wind continue unchanged, there are none formed at all. When the great body of the lower atmosphere is moving from the S.W., this being a more humid wind, the ground is not suffered to get so dry, and the formation of cloud is favoured. It may still happen that even with a S.W. wind there may be no clouds in certain instances; which will be the case, if the moisture, after being evaporated from the soil, and carried upwards by the sun's influence, should pass into a stratum of drier air, capable of retaining it in the invisible form, as when an Easterly or N.Easterly current overlies one from the S.W. in the manner above alluded to (384).

(386.) These clouds, of the cumulus formation, need never cause apprehension, even though the sky be pretty thickly covered with them, except when the weather is only beginning, as it were, to steady itself *after wet,* and the soil still retains great humidity; or when, after the fine weather has already lasted some time, there are indications by the hygrometer of increased humidity in the air. The cumuli will then show a disposition to pass into the more shapeless masses of cumulostratus, and showers, more or less heavy, are liable to occur soon after the hottest time of the day.

(387.) In spring and autumn, when there is a great difference between the temperatures of the night and the day, and occasionally at other times, the ground during fine settled weather is often covered in the evening with a thick stratus. This is, in fact, mainly the subsiding of the vapour raised by the sun's heat during the day, to be retaken up the following morning, forming cumuli or not according to circumstances.

(388.) The weather above described, accompanied by steady winds, may occur at any season of the year. But if it happen in winter, and the wind be in any quarter except S. or S.W., it leads generally to frost. The severity of our winters depends much upon the prevalence of Easterly and Northerly winds in the months of December, January, and February. Ordinarily these winds return only at intervals, giving rise to an alternation of frost and thaw, which is, perhaps, most characteristic of this climate. But if December, as it often is, be mild and open, with winds chiefly from the S.W., the Northerly winds blow with more constancy afterwards, giving rise to more continued frost. In some years the mild weather prevails uninterruptedly through December and January: in this case February and March are liable to be proportionably severe, from the Northerly winds setting in then to restore the balance, and a late winter ensues.

(389.) If frost occur, and continue for any number of days, notwithstanding the wind is in the S.W., we may infer, as in some cases of fine settled weather in summer from that quarter, that there is a N.Easterly current above, which keeps the temperature lower than it would otherwise be under such circumstances. We should have an indication of such being the case, by the barometer standing higher than when the weather is *mild*, with the same S.Westerly wind below.

(390.) We will now suppose that, after a certain

continuance of settled weather, with a prevailing wind from some point in the N.–E. quarter, a change is at hand. The change will generally be longer or shorter about, in proportion to the time for which the drought has lasted. When this has prevailed for several weeks, many days may ensue before the wet comes. During this interval the barometer keeps falling, and the sky gradually gets overcast. It is the barometer, however, which · gives the first sign, its fall attesting the setting-in of an upper current from the S.W., which does not immediately interrupt the serenity of the atmosphere, while the wind below still remains unchanged. After a time, however, the wind veers to E. and S.E.; streaks of cirrus cloud begin to show themselves at a great elevation; these are soon followed by other clouds of the cirrocumulus form, and the sky gradually loses its transparency, until it becomes everywhere of a milky hue. At length the lower current passes on through the S. to S.W., while cirrostratus clouds begin at the same time to form at a lower elevation than the clouds first observed, and to inosculate with them. This state of things having arrived, which is often accompanied by a breeze, more or less considerable, the barometer also having fallen to below its mean height, the rain may be considered as not far off.

(391.) The above is the way in which changes from dry to wet most generally take place. Sometimes, however, the wind seems to veer from N.E. to S.W. more suddenly, though the coming-on of the rain is equally gradual. A good example of this occurred in October 1844, about the middle of which month, a great fall of rain, amounting to more than 2 inches, took place, after a drought of several weeks' continuance. The latter half of September had been very fine and settled, with Easterly winds, and a barometer generally above 30 inches. The first intimation of a change showed itself on the 1st of October, when the wind veered to S.W., the barometer at the same time beginning to fall, and the sky getting overcast, with a *very little* light rain in the afternoon of that day. The wind, with scarcely any variation, continued S.W., *blowing very fresh for six days previous to the rain falling in any quantity*, which did not come decidedly on—though the clouds were very threatening, and a few drops fell occasionally,—till the 14th and 15th of the month. By this time the barometer had fallen to 29 inches. Wet weather then prevailed to the end of the month, the wind chiefly oscillating between S.W. and N.W.

(392.) Long frosts in winter break up much in the same way as long droughts at other periods of the year. When there have been clear skies, with a high barometer and N.Easterly winds, the approach of a thaw is generally first announced by the barometer falling, and the setting-in of an upper current from the S.W., just at which juncture the coldest night usually occurs (49), previous to a thin mist pervading the atmosphere, and the thaw being confirmed.

(393.) But when frosts have set in during changeable weather, and have been accompanied by much snow and a low barometer, the circumstances may be very different. At such times I have often known the wind, at the commencement of the frost, pass from S.W., through W. and N.W., to N.E., but apparently only the lower current, and, when the hard weather broke up, return the same way it came. The wind, however, is never steady at any time during such frosts, and the weather is altogether of a different character from that I have been hitherto speaking of.

(394.) Thaws, especially when the previous frost has been of long continuance, seem seldom to be lasting, except when ushered in by a wind blowing pretty fresh from the S.W., and accompanied by more or less of rain.

(395.) When wet has once fairly set in, let the season of the year be what it will, the weather for a time goes through a succession of changes, which, from their observing no fixed order, it is impossible to characterize in detail.   Still there are differences that may be pointed out according as the rain is due to the intermixing of two winds,—or to a continued flowing-in of a damp S.Westerly current into an atmosphere colder than itself,—or to the mere overgrowth of cumuli, which derive their vapours from the soil of the same locality in which they are precipitated, or of some locality not far distant.

(396.) The wet weather arising from the first of these causes is that which is so characteristic of an English summer, whenever the wind is not steady to the E., or some point in the N.-E. quarter.   The wind in this case is constantly oscillating between S.W. and N.W.; sometimes, perhaps, running up in the latter direction as far as N., and then back again in the other as far as S., but keeping almost entirely to the Western hemisphere. This leads to a frequent alternation of fine and wet, or as it is termed, "changeable weather."   It is curious to notice at such times how the movements of the barometer and thermometer synchronize with the wind's changes,—these movements, however, being in opposite directions,—the thermometer rising when the barometer falls, and falling when the latter rises.   Thus let the wind make an advance Northwards,—the sky, which had been previously clouded, with rain at intervals, becomes clear,—the barometer rises,—the temperature, if it does not fall much by day, at least falls by night, and in any case the air from being more dry is not so oppressively hot to the feelings.   Let now the wind recede to where it was before, and the contrary order of things takes place; though, perhaps, in neither case the effect may show itself immediately, except in respect of the barometer, which in this latter instance falls.   It may continue fine for twelve hours or more, with an increasing, and perhaps sultry temperature.   The wind then gets up, and the sky gradually becomes overcast, ending generally in rain.   Often during the rain, the wind will advance and work on to S.S.W., or even due S., the fall increasing as the wind nears the latter point. The wind seldom, however, remains there long, and as it gradually returns Westward the rain abates, unless, as sometimes happens, the wind makes a sudden leap back to N.W., when the rain falls very heavily indeed for a few minutes, previous to holding up (320).   On the veering of the wind to N.W., a considerable breeze ensues, and the weather soon becomes finer, with light showers, however, occurring occasionally.   In this intermediate state it continues for a longer or shorter period, until the wind steadies itself in that quarter, or, passing on further in the same direction, fixes itself in the N. or N.E., the weather likewise settling,—or else runs back a second time to S.W., and more wet ensues.

(397.) It is this continual struggle between the S.W. and N.W. winds, which, in the early part of the spring, gives rise to the violent storms of hail and rain so frequent in March.   These storms are sometimes attended

by thunder and lightning (371); or, if the temperature be sufficiently low, they consist entirely of snow.

(398.) The wet consequent upon the flowing-in of a
humid S.Westerly wind into a colder atmosphere,—
which was the second cause of rain above mentioned,—
though occurring at other seasons, is more frequent in
the autumn and early winter months, when the temperature of the air at the place where the rain falls is
on the decline.   It often gives rise to dripping weather
of two or three days' continuance.   Sometimes it is an
upper current only which sets in from the S.W., while
the one next the earth, as shown by the vane, is E. or
N.E.   There is then a partial mixing of the two currents
in the plane of contact, occasioning more or less of rain,
according to the degree to which one encounters the
other ; or, if there is not actual precipitation, the colder
current below so far condenses the warmer one above,
as to cause the sky to be uniformly overcast.   In either
of these cases the barometer will be low, though not so
low in the last instance as in the first, when a greater
body of rarefied air is flowing in from the S.W.   The
cloud most prevalent in this kind of weather is the cirrostratus,—altering in character from time to time, but
generally in the form of a diffused sheet, denser perhaps
in some parts of the heavens than others.

(399.) Or the rain may arise in this manner.   After
the prevalence for a certain time of a S.Westerly wind,
with a falling barometer, but without rain, the wind
changes to N. or N.E.   Rain then ensues, from the
vapours which the S.Westerly wind had brought up
being condensed by the colder current from the N.,
while the barometer rises.   The barometer continues to
rise all the time the rain is falling, and it is not till the

N. wind has completely beaten back the S.Westerly one, that the barometer attains that altitude, which speaks to the weather being settled.

(400.) From the two last cases, it will be seen that we may have rain either with a Northerly or a S.Westerly wind, the movements of the barometer and the thermometer being here also in opposite directions, as in the instance of the changeable weather, with the wind oscillating between S.W. and N.W., previously spoken of (396).

(401.) In fact there is no end to the modifications which may take place in the weather, whenever there are two distinct superimposed currents (132). The lower one will often go completely round the compass in the course of the day, after or during wet, when, according to the point in which it rests for a short time, brighter skies at intervals may lead to the idea that it is about to set fine. But the barometer will always tell how far these appearances may be trusted. Unless that rises materially, the rain will be liable to return.

(402.) It is usually during the prevalence of two currents in the air, that, in winter, *white frosts* occur, which, from their soon yielding, have given rise to the saying, "three nights' frost and a thaw." There is often at such times an upper current from the S.W., supplying the air with the moisture, which, when frozen, gives the ground its whiteness; while the lower is in the N.E., occasioning the cold, or revolving, perhaps, slowly round the compass. After a short interval, the lower current rejoins the upper, and the frost disappears.

(403.) The third cause of wet, or that arising from the excessive growth of cumuli, principally in spring and summer, calls for little to be said in addition to what has been already stated on this subject. Rain arising therefrom is necessarily of very local extent, taking the form only of passing showers, which, however, may be heavy for a time, if the air in the region in which the clouds are formed is in a very humid state. And, as the degree of humidity at that elevation cannot be tested by any observations with the hygrometer below, and the barometer is not necessarily affected to any great amount by the changes which may be going on in that stratum of the atmosphere alone, it follows that we have only the appearance of the clouds to guide us in our judgement about the weather, and they must be closely watched to prepare us against rain.

(404.) In all cases of unsettled weather with variable winds, rain is probable, but the quantity that falls is quite uncertain, and in no exact keeping with the descent of the mercury. Thus on the change of the wind to S.W., after it has been some time in the E. or N.E., with clear skies,—or when the wind is oscillating between S.W. and N.W.,—there may be all the appearances above described, when speaking of these two states of weather (390, 396), but the amount of wet actually precipitated will depend mainly upon the humidity of the air in relation to the temperature. And we must take this into account, not merely in respect of the current of air setting in from the quarter to which the wind changes, but in respect of the atmosphere at the place where the observer is situated. Now, though a S.W. wind brings up moisture, and makes the barometer fall, it generally causes likewise an increase of temperature. And the increase of this latter may be so considerable,— especially when combined with the circumstance of a dry air previously surrounding the observer,—as to counterbalance the tendency to wet arising from the increase of humidity. In this case there may be little or no rain: the weather may be still fine, only hotter. Or there may be much wind, with an unsettled state of the sky, but no actual precipitation. This often happens in summer, and the rain we were expecting, we say, "keeps off." The barometer, however, falls just the same; and the circumstance of no rain following surprises persons who are not aware to what the movements of the mercurial column are due.

(405.) There are scarce any rules in meteorology to which there are not exceptions, tending to show how complicated the phenomena of the weather are, and how little we can assert with confidence, from present appearances, what changes are about to take place. As a general rule the barometer falls before rain; but we have seen that, occasionally, it rises with rain (399), as also that it falls without any rain taking place (404). As another general rule, the movements of the barometer and thermometer are in opposite directions, that is to say, "a diminution of temperature is combined with a rise, and an increase with a fall of the barometer." In fact the pressure of the atmosphere, from its being so closely connected with the changes of temperature, has been compared by Kämtz to a *differential thermometer**. But even this rule sometimes fails us; and in the instance I am about to adduce, both rules will be found in fault at once: not only did the barometer fall to the amount of an inch without any rain immediately following, but there was a simultaneous fall of the temperature equally remarkable.

* See *Edinb. New Phil. Journ.* vol. xxvi. p. 264.

The instance in question occurred in the last week of September, 1843. Very fine settled weather had prevailed all the previous part of that month, with a high temperature, the daily maximum seldom falling below 70°, and often being as high as 75°,—on one occasion 77°·5. During this period the barometer had been generally high, and on the 23rd had reached the elevation of 30·541 inches. The following day a fall of the barometer commenced, which continued gradually till the 27th, when the mercury stood at scarcely more than 29·5 inches. But with the exception of a very little light rain on the 25th, and again on the 30th, this last occasion being three days after the barometer had begun to rise again, there was no rain till the following month, when the weather became changeable, and from October 7th to October 12th very wet. The only change observed at the time of the barometer beginning to fall, on the 24th of September, was a shift of the wind from N.E. to N.W., through S.W. and W., with misty clouds, and a lowering of the temperature. On the 25th the wind passed to N., where it continued several days, with much cloudiness and a still greater fall of the temperature, the mean daily maximum for the last week of that month being only 57°, whereas the same mean for the first three weeks of September was 72°.

It turned out in fact that the above great fall of the barometer was indicative of a general break-up of the fine summer weather which had previously prevailed, but not of rain to any amount, at least immediately.

(406.) There is often difficulty in accounting for insulated phenomena which show a departure from ordinary laws; but in meteorology the difficulty is

mainly derived from the want of simultaneous observations at remote places.   Did we know exactly all
that was going on elsewhere, both above and around
us, we should probably, in many instances, find an easy
explanation of the disturbed relations between the several conditions of atmosphere that may appear in a particular locality.   Thus, with reference to the above
case, in which there was shown to be a great fall of the
barometer, accompanied by a considerable fall of the
temperature as well, contrary to what usually takes place,
there was, in all probability, an equally considerable
rise of both somewhere else at the same time.   We
must remember, further, the disadvantage we are under,
in attempting to explain such an occurrence, from the
circumstance of the thermometer merely giving us the
temperature of the air to the height of a few feet above
the ground, at the place where the observation is made,
while the barometer at the same place is affected by
the pressure of the whole atmosphere to its extreme
limits.   Clearly, to judge in such a case whether there
is any real deviation from the ordinarily observed rule,
viz. that the movements of the barometer and thermometer are in opposite directions, we should know the
mean temperature of the whole column of air above the
place of observation, just as the barometer tells us its
whole pressure.   For although the former " depends, in
a general way, on the indication afforded by the thermometer at the surface, yet it can easily happen in particular cases, that the mean temperature of the whole
mass of air may diminish, while it becomes higher at
the surface, and *vice versâ* *."

* Kämtz in *l. c.*, to which the reader is referred for further
remarks in explanation of the above anomaly.

(407.) In the foregoing part of this chapter, it has been attempted to point out some of the features attending the principal states of weather, characteristic of the climate in that part of England to which the observations refer. But there is one chief circumstance upon which most of them depend, and to which, in concluding this part of the subject, as a general rule, it may be well to call attention. I allude to the direction in which the wind moves being ordinarily with the sun, or like the hands of a watch. If the wind change from N.E. to S.W., its passage is through the E., S.E., and S.; if it change from S.W. to N.E., its passage is through the W., N.W., and N. At least if it move in the opposite direction, it is only during unsettled weather, when it is constantly oscillating, and we may soon reckon upon its return. This "law of rotation," which has been strongly advocated by Dove, and before alluded to (108, 109), entails a certain fixed succession of the phenomena connected with pressure, temperature, and humidity, as the wind progresses onwards. The N.E. wind being the coldest and the driest, while the S.W. is the warmest and the moistest,—the intermediate winds being in regular gradation,—it follows that as the wind shifts from the former to the latter quarter, through the E., S.E., and S., it is necessarily attended by a falling barometer and a rising thermometer, together with an increase of moisture in the atmosphere. As it pursues its course further, to complete the circuit, or shifts from the S.W. to N.E., through the W., N.W., and N., this order of phenomena is inverted : the barometer rises, the thermometer falls, and the air becomes drier.

(408.) The various changes of weather, complicated

though they be in many instances, seem mainly in accordance with this order of sequence of the winds, and
the altered conditions of the atmosphere brought about
from time to time by their agency. There are also suggested to us certain differences, which, under the influence of the above law, serve to distinguish the weather
and its changes, according as the wind at the time is
passing to or from the S.W. Thus, in consequence of
the amount of vapour in the air increasing as the wind
advances from N.E. to S.W.,—and decreasing in the
passage of the wind from S.W. to N.E.,—the temperature likewise varying in the same way,—it follows that,
while the rain which occurs with a Southerly wind has
its origin in an excess of moisture in the atmosphere,
—that which occurs with a Northerly wind is due, not
to the vapour being in any excess, absolutely considered,
but to its condensation by cold. Another circumstance
that may be pointed out respects the different way in
which changes of weather, generally, are brought about,
according as they take place in the E. or in the W.
When the wind is in the E., about to pass Southward, a warmer current succeeds a colder one: when
the wind is in the W., and passing Northward, a
colder current succeeds a warmer. But in the former
case, the warmer current from the S.W., being the
lighter of the two, shows itself first *above*, and is only
very gradually substituted for the colder one below;
whereas in the latter the colder, being the heavier, is
introduced from *below*, and gradually extends itself
upward. This occasions changes of weather to come
about much more slowly in the E. than in the W. In
the former instance, the barometer falls so soon as ever
the S.W. current begins to exert its influence in the

upper regions of the atmosphere, but it is not generally
till some little time afterwards that the wind changes
below, or that the air attains that degree of humidity
that leads to precipitation in rain. In the latter, the
shift of the wind from W. to N.W., the rise of the
mercury, and the fall of the temperature, are almost
simultaneous.; and if there be much moisture in the
air at the time of the wind changing, a shower occurs
almost immediately.

(409.) Persons must have often noticed the above
difference. When the change is in the E., it is ordinarily a change from dry to wet; but we see the clouds
forming at a great elevation, in consequence of the introduction of the warmer current above, long before
the rain actually descends. When the change is in
the W., it is, on the other hand, ordinarily a change
from wet to dry. Clouds, it is true, form also immediately, and, as just stated, there may be a copious
precipitation due to the sudden intermixing of two
currents of different temperatures; or a succession of
such showers, for a day or more,—less and less rain,
however, falling on each occasion, as the Northerly wind
more and more supplants the current from the S.W.
But after a time the air is cleared of all its superabundant vapour, and if the wind continue to pursue
its course to N.E., the barometer likewise gradually
rising to the mean height of the mercurial column belonging to that wind, we may then infer that the weather
is again settling fine for a season.

(410.) The above order of phenomena likewise explains why, in winter, such precipitations of rain in the
W. frequently turn to snow : it is from the air getting
colder, as the wind works Northward ; whereas, on the

breaking-up of a frost, with the wind in the E., it first snows, before the warm air above has got thoroughly mixed with the cold below, and then rains. In the one case the barometer rises, in the other it falls. "Snow," therefore, with a "rising barometer indicates a more rigorous cold; with a falling one, a milder temperature *."

* Kämtz, p. 312, from whom some assistance has been derived above.

CHAPTER VII.

OF WEATHER PROGNOSTICATIONS.

(411.) SOME persons seem to think that meteorology
has no more legitimate object than to afford predictions
about the weather. They attach little value to any
investigation of the general principles upon which the
science is based, and the laws which regulate the distribution of heat and moisture throughout the globe,
if such investigation lead to no practical advantage in
the way alluded to. Or if they do not take quite so
narrow a view of the subject, they at least anticipate
that the science will one day attain to such perfection,
as to reveal the future, in respect of the weather and
the seasons, with almost as much accuracy as it records
the past. The pains and labour bestowed in these days
upon Meteorological Registers,—kept too in so many
different places all over the world,—seem to them to
encourage some hope of this kind. If, say they, the
weather is under the influence of any definite laws at
all, the knowledge of these laws, as determined by what
has been, will surely lead to the discovery of what is to
be. The phenomena which the laws bring about, will,
if the latter be constant, return from time to time in
some given order, conformable to that in which they
have appeared before.

(412.) The question whether such expectation is ever likely to be realized, is at least worthy of consideration; for doubtless it would be a great matter, if we could always reckon upon the weather of any particular day or week for which our schemes of pleasure or business were set, with as much certainty as we reckon upon the changes of the heavenly bodies predicted by astronomers. If the farmer, by consulting his almanac, could fix the exact time for sowing his seed, or cutting his hay, so that in one case he might ensure rain following shortly afterwards to bring the seed up,—in the other, secure a dry period for getting in his harvest,—he would have a great advantage over those who had no such authority to guide them in their operations.

(413.) But are we ever likely to arrive at this? To judge, indeed, by the weather almanacs, which yearly make their appearance in no small number, one might suppose that the science had already made sufficient advances to warrant the predictions of those who set themselves up to be prophets in this matter. But it is hardly necessary to warn the public against placing the slightest confidence in these publications, which have been so often exposed. In some instances these almanacs have acquired notoriety for a time, by a few happy guesses about the weather, which have come right by a mere coincidence; but in the long run, if any one will take the trouble to compare them throughout with what really occurs, their predictions will be found just as often wrong as right, showing that they are grounded upon no trustworthy principles.

(414.) Some, indeed, pretend to base their foreknowledge of the weather upon the foreknown changes of the heavenly bodies above alluded to. They claim to be listened to on the ground that, the weather being under the influence of the moon and planets, and altering from time to time as these bodies alter their positions in respect of the earth and each other, we may safely draw our inferences about the former, from knowing the exact places of the latter, on any particular day or month we may have in view. But greater names than any which this class of meteorologists can boast of have utterly discouraged all such theories. Arago for one, in reference to the common notion of the weather being affected by the moon or comets, has expressed his belief that, if the latter have any influence at all, that influence is so small as to be almost inappreciable, and that consequently "the predictions of the weather can never be a branch of astronomy properly so called*."

(415.) With respect to the moon, as distinguished from the other heavenly bodies, it may be proper, however, to say a few words in addition. From its greater proximity to the earth, this luminary might with more reason be supposed to have some influence on the weather. And the question whether it has or not, has engaged the attention of a few really scientific men, who have considered it in different points of view. By some, indeed, it has been thought hardly credible, that the popular belief—"that changes of weather oftener take place about the full and new moon, and about the quadratures, than at other times,"—should have come to be so generally prevalent, unless some connexion between the moon and the weather really existed. One meteorologist, of well-established reputation in this country, has even gone so far as to pronounce such connexion

* *Edinb. New Phil. Journ.* vol. xli. p. 3.

to be "a fact founded on long observation*." But when we come to inquire more about this supposed "fact," and to look into the evidence for and against it, supplied by those who have taken the trouble carefully to investigate the subject, we find our doubts only increasing instead of being dispelled, until we are left as much in the dark as ever about the truth.

(416.) Of the observers who have gone into the question, some have endeavoured to ascertain, whether there is any excess of rain corresponding with the periods of any of the moon's phases. Others have sought to trace the influence of the moon upon the barometer : others, again, the effect of the same upon the winds, temperature, &c. But from none of these inquiries, to some of which I will refer more particularly,—however inviting in themselves, and however plausible the results may seem at first in some cases, individually considered,—do we derive any facts having the slightest practical bearing upon our speculations about the weather. The fact rather is, that the conclusions arrived at are often so at variance with each other, as entirely to shake our confidence in the belief that the moon exercises any influence upon the weather at all.

(417.) Thus with respect to the rain, one observer, M. Flaugergues, taking the number of rainy days coincident with the days of the moon's phases, during a period of nineteen years, found that the number was at a maximum at the first quarter, and at a minimum at the last quarter†.

Another observer, Schubler, found, as the result of twenty-eight years' observations, that the number of

* Forster (*Atmospheric Researches*, p. 156).
† *Edinb. New Phil. Journ.* 1828, p. 318.

rainy days was at a maximum on the day of full moon, and at a minimum at the last quarter*.

Dr. Marcet, again, was led to conclude, from the observations of thirty-four years, that the rainy days were, as in the last instance, at a maximum at the time of full moon, but at a minimum at the *first* quarter, or just the reverse of M. Flaugergues. There is so little difference, however, in the actual number of rainy days at the four several periods, in Dr. Marcet's observations, that it is evident not the slightest importance is to be attached to the difference that exists†.

(418.) There is the same uncertainty about the moon's influence on the barometer. Some observers think they can clearly detect it, by taking the average height of the mercurial column at the several periods of the moon's phases, while others fail to do so in any satisfactory manner. Howard‡ and Daniell§ were both of opinion that the barometer attained a greater elevation, on an average, at the time of the first and third quarters, than at the times of new and full moon. Flaugergues thought that the mean height of the barometer was greater at the new moon than at the full, and greater in the last quarter than in the first, the greatest mean height (coinciding with the least number of rainy days as given above) being in the last quarter, and the lowest mean height (coinciding with the greatest number of rainy days) being in the first quarter.

Taking the mean of my own barometric observations on the several days of the moon's phases during a period of nineteen years, I find, like Flaugergues, the mean

* *Edinb. New Phil. Journ.* vol. xviii. p. 359.　† *Id. loc. cit.*
‡ *Climate of London,* vol. i. p. 156.
§ *Meteorological Essays,* p. 137.

height of the barometer greater at the time of the new moon than at the full, and greater in the last quarter than in the first; but the greatest mean of all is at the time of new moon, and the lowest at the full, and not in the last and first quarters respectively as he makes them.

Mr. Lubbock, on the other hand, in an article devoted to the discussion of the meteorological observations made formerly at the Apartments of the Royal Society, with a view "to ascertain whether any variation takes place in the dew-point, and also in the thermometer and barometer, corresponding to the moon's age," arrived at the conclusion that if *any lunar* inequality exists, which he was unable satisfactorily to prove, "it must be very minute and inadequate to afford a secure basis for predictions of the weather*."

(419.) Dr. Marcet, in the paper already referred to, considered this question of the moon's influence on the weather, in reference to another point. In order to test the truth of the popular notion, as to changes of weather (limiting the expression to changes "from clear weather to rain, or from rain to clear weather") occurring more frequently "on the four principal days of the lunar phases than on common days," he examined a register of the weather kept at Geneva for thirty-four years. The results which he obtained, and which it is not necessary to give in detail, seemed, "upon the whole, to lend some support to the vulgar opinion of the influence of new and full moon, but none whatever to any special influence of the first and third quarters." But then this support was very slight; and, as a set-off against it, Mr. Glaisher of the Greenwich Observatory has declared,

* "On the Supposed Influence of the Moon upon the Weather," *Companion to the Almanac* for 1839, p. 5.

as the result of the meteorological observations made
at that establishment since the year 1840, that "changes
of weather have been found to be as frequent at every
age of the moon, as when she has been seven, fourteen,
twenty-one, or twenty-eight days old; therefore she
cannot have had the slightest influence over any of
them*."

(420.) The above are a few of the conclusions arrived
at by different observers in the inquiry under consideration : it is unnecessary to allude to others. When we
look to the contradictory nature of those just referred
to, it is quite evident that we ought, for the present at
least, to dismiss from our minds all idea about the moon
influencing the weather at all. There may be, as many
think there are, tides in the atmosphere, in like manner
as there are tides in the ocean, caused by the moon's
attraction, and it is a subject which may well engage
the attention of astronomers and professed meteorologists. Possibly it may one day be found that the
moon's gravitation is a necessary element to be taken
into consideration with the other laws that regulate the
movements of the great aërial ocean around us. But
whether so or not, it is clear that this matter may be
set wholly aside in all our reasonings about the weather.
If the moon have any influence on the weather, that
influence is so slight that it must necessarily be constantly overruled and counteracted by other influences
having much greater weight. Assuming its existence,
it is only to be detected by taking the average results
obtained from observations carried on for a great length
of time : this would not be, if a coincidence between the

* Quoted by Mr. Lowe in his *Prognostications of the Weather*,
p. 27.

moon's changes and the changes of the weather took place as often as is supposed.

(421.) But it may be asked, how is it that this notion respecting the moon's influence on the weather has so generally prevailed, from times far back, if it be not entitled to credit? It may be difficult to answer this question to the satisfaction of those who put it, but it is equally difficult to trace to their origin many other popular fallacies which have crept into existence, and obstinately maintained their ground, long after having been exposed by the researches of modern days. It does not necessarily follow that an idea, because generally prevalent, has truth for its foundation. With respect to the idea in question, we are quite sure of this, —that the meteorological observations carried on at the present day, are far more accurate and extensive than those made in old times. If, therefore, there was any truth in the opinion that the moon influenced the weather, this truth, instead of being rendered more doubtful by modern investigations, ought to come out the more clearly. That it does not so come out is strong evidence against the opinion being trustworthy. We know too how, at one time, the heavenly bodies were thought to have a powerful influence over the destinies of men, the affairs of common life, and various other matters ; and this supposed influence of the moon on the weather may be the last lingering remains of a superstition, which in other shapes has been now nearly exploded. Those who still hold to such an idea, are generally persons, who, like farmers, are naturally much interested in knowing what the weather is likely to be, but who seldom or never make any observations respecting it, except those to which they are accustomed by habit, or inclined by prejudice. And having been once led to entertain the notion, that changes from wet to dry and the contrary (which is all they think about) generally happen either at the full or new moon, they are induced to look for such changes at those times when it is of great importance to their interests to secure the exact weather they want. If, in a few cases, the expected coincidence comes about, they are naturally much strengthened in their views respecting the moon's influence in this matter. But they seldom take any account of those cases in which the coincidence fails, and if it be not a season in which it is of so much importance to them what the weather may be, the subject escapes their observation altogether.

(422.) While some have entertained the idea that, the weather being under the influence of the moon, it is possible to prognosticate to a certain extent, what weather is likely to prevail, according to the moon's place in the heavens; others have been led to think that there is a certain cycle of changes, through which the weather passes, extending over a given number of years, and that when the cycle is run out, the same changes recur in the same order. This theory, which views the weather rather in its collective character, as regards seasons, than in its less important changes from day to day, is by no means new. It dates at least as far back as Bacon, who himself had some notion of there being such a cycle of five-and-thirty-years, within which the round of variations took place*.

---

* "There is a toy, which I have heard, and I would not have it given over, but waited upon a little. They say it is observed in the Low Countries (I know not in what part), that every five-and-thirty years the same kind and suit of years and

(423.) In modern times this notion has been revived by some meteorologists, especially by Howard, who, in a communication made to the British Association in 1842*, as well as in a tract separately published the same year†, has shown an inclination to adopt eighteen years as the length of the cycle, in respect at least of the seasons of this country.

(424.) Howard found that during such a cycle, commencing in 1824 and terminating in 1841, "the seasons at Ackworth in the West Riding of Yorkshire, went through their extreme changes of wet and dry, of warmth and coldness; returning at the end to the same (or nearly the same) state again." The first nine years of this series, " as comprehending seven years of a temperature above the mean, he termed *The Long Summer*; and the latter nine, having seven years below the mean, *The Long Winter* of these Islands; not that the effects are so confined, the like seasons prevailing to a great extent, in like succession, in the European continent also."

(425.) It does not appear, however, that these two series of years, taken each as a whole, were characterized by any differences of wet and dry, corresponding to the above differences of temperature, the amount of rain being stated by Howard to have been nearly the same in each. And it is surely questionable whether there are weathers comes about again; as great frosts, great wet, great droughts, warm winters, summers with little heat, and the like, and they call it the prime: it is a thing I do the rather mention, because computing backwards, I have found some concurrence. —*Essay LVIII.* "Of Vicissitude of Things."

* *Rep. Brit. Assoc.* 1842, (Trans. of the Sects.) p. 24.

† *A Cycle of Eighteen Years in the Seasons of Britain; deduced from meteorological observations made at Ackworth, in the West Riding of Yorkshire, from 1824 to 1841, &c.* 8vo. Lond. 1842.

any real grounds for the theory he has proposed. We may find the weather conforming to a given order of changes in two or three consecutive instances, but it would be extremely hazardous to affirm that such an alternation of series of warm and cold years would be found always to recur. It would at least require a very lengthened period of time to pass by, before we could obtain sufficient proof of the correctness of the theory, to allow of our generalizing in this way.

(426.) The very circumstance of different observers differing in their opinions respecting the length of the cycle, within which the order of changes takes place, is itself unfavourable to the idea of there being any fixed cycle at all. And in truth, a cycle of eighteen years, as imagined by Howard, seems hardly reconcileable with facts. On referring to the temperatures of a long series of years, as deduced by Mr. Glaisher from observations made at Greenwich, and published in the ' Philosophical Transactions,' these temperatures present no such regular alternations of hot and cold, as to lend any support to the theory in question, though Mr. Glaisher thinks it appears that, during the seventy-nine years, over which the above observations extend, "the temperature of the climate has increased*." Howard's long summer reached from 1824 to 1832, and had a mean temperature (at Greenwich) of 48°·9. His long winter reached from 1833 to 1841, and had a mean temperature (at Greenwich) of 47°·2. But if we work backwards from 1824, by the help of Mr. Glaisher's Table above alluded to, and take the temperatures of the three preceding series, of nine years' each, the result is as follows :—

* *Phil. Trans.* 1850, p. 585.

Mean of nine (cold) years (1797–1805)—48°·2
.   „   „ (warm)   „   (1806–1814)—48°·8
    „   „ (cold)   „   (1815–1823)—48°·7

I have called the above series cold or warm, as they *would be*, if they harmonized with Howard's theory. But in fact it will be seen that, except in the case of his "long winter" from 1833 to 1841, when the mean temperature fell to 47°·2, there is but little difference in the temperatures of any of the series of years so parcelled-out. Moreover, there is a greater difference between the first and third series, which, according to Howard's views, ought to be both cold ones, than between the second and third, which ought to be, the latter of the two a cold series, and the former a warm one.

(427.) But setting aside cycles, no less than the supposed influence of the moon, the truth is, that the more the science of Meteorology advances, the less hope there seems to be of our ever being able to foretell the weather with any certainty. The question may be thought to assume a different aspect, according as we desire to determine beforehand whole seasons, or to know what the weather may be a few days only in advance. But Kämtz, a high authority, who has treated of this matter, has dwelt on the difficulties by which both these inquiries are surrounded. With regard to the first, he has observed, that, "owing to the intimate connexion existing between all the different parts of the atmosphere, an approximate solution would only be possible if we were acquainted with the actual state of the weather over a large portion of the earth. As this is not the case, the solution of the problem would be impossible, if the barometer did not afford us some feeble hints on the subject, by means of its property of acting as a differential thermometer. When it stands unusually low, and at the same time exhibits great disturbance, we may thence conclude that other regions are very cold, and that we shall not only very soon receive back a portion of the air from them, but also, that the weather will for some time follow an unusual course. But more cannot be said with certainty, as we cannot know at the moment of observation, whether this great cold has its position in the interior of Siberia or of America. If the first is the case, then we may soon expect N.E. winds; whereas W. winds come from America, and these bring us in winter a moist and warm air. It is thus possible, that the same phenomena may be combined with perfectly opposite kinds of weather."

(428.) If it is " difficult to predict satisfactorily the character of whole seasons," Kämtz thinks there is greater difficulty in indicating " how the weather is to be at an interval of a few hours. In an investigation of this kind, we must not merely keep before our eyes the character of the whole season, but an accurate knowledge of the whole atmosphere above us is also requisite, which, from the very nature of things, is perfectly impracticable in reference to temperature and moisture. It is true that the exertions of travellers have shown us how these relations change, as we proceed from the lower to the upper strata of the atmosphere; but these investigations relate to the mean state of the atmosphere, and very important errors are possible when they are applied to particular cases. We know (to adduce only one example) that during a certain mean state of the hygrometer, rain generally takes place; the barometer sinks at the same time, and the

probability of the precipitation becomes greater, especially if the sky begins to be obscured by clouds.   But
in order to predict with certainty if it will rain or clear
up, a knowledge of the temperature of the upper region
is requisite, and, as this is wanting, there must always
be a great degree of uncertainty in our prognostications.
Supposing the temperature at a height of 10,000 feet
to be some degrees lower than usual, a great precipitation would be the consequence; whereas, if the temperature should rise an equal number of degrees, the sky
would clear up with rapidity*."

(429.) These last remarks by Professor Kämtz seem
especially referable to that state of doubtful weather, of
which we have so much in this climate, when we never
can determine what the result will be.   But even when
the weather is to all appearance quite settled, changes
from dry to wet sometimes take place so suddenly, that
it is hardly possible to foresee them, especially if they
come on in the night.   After a fine day, continuing
clear till midnight, with the barometer at 10 P.M.
standing at nearly $30\frac{1}{2}$ inches, and the moon shining
quite bright, on rising the next morning, I have found
rain falling steadily from a thickly clouded sky, and
continuing to fall the whole day.   These sudden changes
have occurred in the autumn, when the temperature has
been low.   They are probably more frequent at that
season than in summer, the atmosphere being then in a
more humid state, and slighter fluctuations of temperature and pressure being sufficient to cause a precipitation.

(430.) Of course, in such a case as that just mentioned, there is some change of wind, or shifting of

* "Remarks on the more important Atmospherical Phenomena."—*Edinb. New Phil. Journ.* vol. xxvi. p. 272.

currents, either in the higher regions of the atmosphere or next the earth, which the barometer will
indicate, before the rain takes place. But unless the
instrument be very narrowly watched, which it seldom
is during the night, the rain is apt to take us by surprise, at least when we judge of the next day's weather,
from what the latter may have been, even to a late
hour, the day previous.

(431.) As we are thus sometimes surprised by rain
when we least expect it, so at other times, when there
is every indication of a change being at hand, we are
occasionally deceived by its not taking effect. This is
due to the weather running back. When speaking of
the general conditions under which a change from dry
to wet ordinarily occurs, I alluded to the different appearances by which we are led to anticipate it (390).
The wind veers, the barometer falls, and the temperature rises, while the sky becomes partially overcast.
And if this state of things continue, rain, no doubt,
is extremely probable. But then, in some instances,
the wind, after having made a certain advance towards
the S.W., runs back to its former quarter, the other
indications of rain gradually subside, and the same fine
weather returns which prevailed before. It is impossible to tell beforehand whether this will be so or not.
All we can say is, that there is a tendency to wet, which
may, or may not, come about. And even supposing
that the wind pursues its course, and that the above
appearances continue, it does not necessarily follow
that rain will be the result. Allusion has been made
to cases, in which there have been considerable falls of
the barometer, accompanied by a change of wind, but
not by rain, at least to any amount (405). We may

safely infer that *some* alteration of weather will take place, but this may simply show itself in an increase of temperature, or, as in one instance mentioned, a *decrease* of the same,—or in a strong wind,—or in a sky previously clear becoming more or less overcast.

(432.) But there are other difficulties besides those already stated, which stand in the way of weather prognostications, and which from their very nature are insuperable. I allude to causes, often exercising a decided influence on the character of whole seasons, which, being of accidental occurrence, can never be foreseen. It is obvious that any predictions we make about the weather must be founded on general laws. We may never indeed arrive at such a complete knowledge of even these laws, from their complexity, as to be able to avail ourselves of the help they would afford us in judging of the weather, if properly understood. To say the least, we are not masters of them at present. For though we may fix with some degree of precision the mean conditions of any climate, as determined by the observations of a long term of years, we cannot explain the annual fluctuations of temperature and humidity, so as to say why certain years are so much hotter or wetter than others in a given series. Nevertheless we still believe such fluctuations to be under the control of *some* laws, and which operate in restraining the former within certain limits. But are we sure that the conditions of a climate, as hitherto determined, will remain always the same from year to year, or that future seasons will not deviate widely from the character of those past, owing to circumstances quite independent of any such general laws as have been just supposed ? This is the point to be considered ; and many such circumstances, which must be borne in recollection, when we form our judgment of the weather that is to be, have been suggested by different observers*.

(433.) Thus Arago has alluded to the unusual number of icebergs, which, in some years, have been detached, by causes purely accidental, from the general mass of ice in the Polar seas, and borne in an unmelted state into lower latitudes. Large fields of floating ice are said to have been encountered by ships in latitudes as low as between forty and fifty degrees North, in the months of March, April, and even May. These, when they approach within a certain distance of our shores, cannot fail, by the chilling atmosphere with which they are surrounded, to have an unfavourable influence on the spring and summer of the western parts of Europe.

(434.) Mr. Fairbairn, in a communication to the British Association in 1842, has also spoken of this circumstance. On the occasion alluded to, he mentioned the fact of the unusual presence of ice in the Atlantic Ocean, as a probable cause of the low temperature of the summers of 1838, 1839, 1840, 1841. He added, that " about half a century since, icebergs were a phenomenon in the Atlantic not visible until the autumn ; now they increase annually in numbers, and

* Professor Forbes has remarked, that "there are several causes which may tend to change the amount of rain on a particular spot without forming part of any general law: among such changes will be found the destruction or the planting of forests, the enclosure and drainage of land, and the increase of habitations."—*Report Brit. Assoc.* 1832, p. 251. See also Arago's memoir on this subject in the *Annuaire pour l'an* 1846, of which a translation will be found in the *Edinb. New Phil. Journ.* vol. xli. p. 3. From this last author, some of the remarks which follow above have been in part borrowed.

appear as early as the month of March. This augmentation of ice in the Atlantic was referred by the author to the increase of warmth on the North American continent, consequent on the cutting down of the forests and the extension of civilization. By this means the ice is detached from the circumpolar bays and rivers at an earlier period, and floats further south before being melted than heretofore*."

(435.) As the summers of our climate are thus liable to be rendered colder than usual, by the occasional presence of icebergs in the Atlantic, which it is clearly impossible to foresee; so there is another cause, the Gulf-stream, which sometimes renders our winters much milder than usual, and which is equally beyond our foreknowledge. From the high temperature of its waters, and from the course it takes, this stream must always have considerable influence in mitigating the cold of this country†. In certain seasons, however, whatever may be the cause, it would seem to have a much further extension in the direction of the British Islands than in others. This has been suggested by General Sabine as the cause of the extreme mildness of the

---

* *Rep. of Brit. Assoc.* 1842 (Sect. Proceed.) p. 26.

† "The maximum temperature of the Gulf-stream is 86°, or about 9° above the ocean temperature due the latitude. Increasing its latitude 10°, it loses but 2° of temperature; and, after having run three thousand miles towards the North, it still preserves, even in winter, the heat of summer. With this temperature it crosses the 40th degree of North latitude, and there, overflowing its liquid banks, it spreads itself out for thousands of square leagues over the cold waters around, and covers the ocean with a mantle of warmth that serves so much to mitigate in Europe the rigours of winter."—Maury (*Physical Geography of the Sea*, Ch. II. Sect. 63).

winter of 1821–22, in which year, he observes, the
Gulf-stream, "instead of terminating, as it usually does,
about the meridian of the Azores, extended to the coast
of Europe." The same author thought that the mild
winter of 1845–46 might have been due to a similar
cause, though at the time of writing he had no evidence
to prove or disprove such an opinion. But then he
adds, that "strange as it may appear, this remarkable
phenomenon may take place in any year without our
having other knowledge of it than by its effects, although
it occurs at so short a distance from our ports, from
whence so many hundred vessels are continually cross-
ing and recrossing the part of the ocean where a few
simple observations with the thermometer would serve
to make it known*."

(436.) Another circumstance of purely accidental
origin, but having a most prejudicial effect on our summers, would be the recurrence at any time of that
"peculiar haze, or smoky fog," as White of Selborne
terms it, which hung in the air for weeks together
during the summer of 1783, and to which allusion has
been before made (280). This phenomenon, which, in
that instance, was undoubtedly occasioned by the presence of some foreign matter in the air, probably the
dust of a volcanic eruption†, caused an obscuration of
the sun, according to Arago, for nearly two months.
Such an obstruction of the solar rays could not take place
without lowering the mean temperature of any country

* See his paper "On the cause of remarkably mild winters,
which occasionally occur in England."—*Phil. Mag.* 1846, p. 317.

† The dust arising from a volcanic eruption may be carried
by currents many hundreds of miles from the spot in which the
eruption takes place.

in which it occurred for the period while it lasted. But
this would "considerably disturb the course of the seasons." For the character of one season must in some
measure depend upon that of the season preceding, and
we are unable to say to what extent the effects of a phenomenon of the above nature might not reach.

(437.) There are other accidents of a more local character, brought about by human agency, which might
influence the weather in particular districts, and which,
if carried to any great extent, might lead in time to an
alteration of climate. Such are the clearing of extensive forests, or the drainage of marshes. Both these
would have the effect of rendering a country less
humid, and diminishing, probably, the quantity of rain.
Thus it has been asserted that "a progressive diminution of rain has taken place in the South of France, in
consequence of the destruction of mountain woods*."
Nor would the effect stop here. The different phenomena of the atmosphere are so connected, and so dependent on each other, that a change induced in any
one of the conditions of a given climate must necessarily have an indirect influence on all the other conditions. If, in either of the above ways, we lessen the
humidity of the soil, there will be less vapour raised from
it to be condensed in cloud or mist, and consequently
there will be more sun. This must lead to alterations
of temperature. But any alterations of this last must
cause currents in the atmosphere, which again may create
further disturbances in the existing state of things, that
we cannot follow out in all their consequences†.

* *Encycl. Metrop.*, Art. "Meteorology," Sect. 584.

† Maury, in his work on *The Physical Geography of the Sea*,
has given an animated account of the extent to which we may

(438.) Nor can it be doubted that the changes effected in this way in one country would reach to others, where there might be no suspicions of the cause to which they are due. While much of the weather of a given district is dependent upon local circumstances, a great deal of it is brought up from distant regions through the agency of the winds. It is on this ground that Kämtz, as before stated, observes that, in order to determine what is to be the character of the next season, we ought to know what the weather is, at the time of inquiry, at all other places. The same peculiarities of

conceive the currents and climates of the ocean to be influenced by the secretions of mollusks and corallines, though in themselves nearly stationary, in abstracting from the sea-water solid matter for their cells. Supposing for a moment all other agents in nature capable of disturbing the equilibrium of the ocean withdrawn, and its waters to be everywhere at rest, and that in this state of things "a single mollusk or coralline commences his secretions,"—he observes, "In that act, this animal has destroyed the equilibrium of the whole ocean, for the specific gravity of that portion of water from which this solid matter has been abstracted, is altered. Having lost a portion of its solid contents, it has become specifically lighter than it was before; it must, therefore, give place to the pressure which the heavier water exerts to push it aside and to occupy its place, and it must consequently travel about and mingle with the waters of the other parts of the ocean until its proportion of solid matter is returned to it, and until it attains the exact degree of specific gravity due to sea-water generally."

Viewing then, collectively, the effect produced by the agency of marine animals generally in this respect, he says that "they have power to put the whole sea in motion, from the equator to the poles, and from top to bottom."—Ch. IX. Sects. 545-548.

The same reasoning seems applicable to any disturbances in the *atmosphere* from local causes, as above supposed, however slight in themselves these causes may be.

weather are rarely common to very widely extended regions. Extremes in one country are met by opposite
extremes in another*. It is in this way that the general
equilibrium is maintained; and the irregularities which
occur in one quarter of the globe are after a time transmitted to some other. If then the winds, by which such
transfer is at any time effected, traverse in their passage
to us, countries in which there has been any change of
atmospheric conditions from the causes above alluded
to, the weather brought to us by these winds would be
surely, to a greater or less extent, different from what
it would have been, if no such change had taken place.

(439.) Thus we see how all our anticipations of the
coming weather, as founded upon general laws, and the
existing state of things, are liable to be set aside by
contingencies, which no human foresight can reach.
We know not, except in the most imperfect manner,
what is going on above and around us, and if we did,
we know not what may any day happen to overturn the
conclusions to which theory might lead us. The accidental occurrences which have been alluded to above,
as having a more or less decided influence on the
weather, are by no means all that might be indicated.
Many others have been pointed out by Arago, in the
memoir before referred to, but which it is unnecessary
to particularize. Suffice it to say, that no one who has
the least acquaintance with the complexity of the laws
by which the phenomena of our atmosphere are ordinarily controlled, and the degree to which these laws may

* Dove observes that "a mild European winter is made up
for by a cold one in America or Asia." And he adduces, as
instances, several particular seasons in which this has been the
case. See *Edinb. New Phil. Journ.* vol. liv. p. 225.

be still further entangled by foreign influences, will
pretend to the character of a weather prophet, in regard
at least of the weather for any considerable time in advance.   If it is ever possible, as perhaps it sometimes
is, to say beforehand what the weather will be tomorrow,
or a few hours hence, it is only in those cases in which
the weather has been for several days past in that settled
state, in which we may safely assume that a change to
wet will not occur without those premonitory symptoms
that have been already described.   In all other cases
there must be great uncertainty.   We may have grounds
for forming a reasonable conjecture as to which way
things will go, but we can never predict the result with
any confidence.   The few rules that can be laid down
for our guidance in this matter, and to which I shall
shortly advert, must be regarded only as of the most
general character, and as all liable occasionally to mislead.   None should be exclusively trusted to, but they
must be compared among themselves, and our judgment
determined by as many as seem to point one way in the
prognostications they afford.

(440.) One thing is certain, that to whatever extent
it may or may not be possible to foretell the weather, those
alone, in general, can form a right judgment who are
possessed of good meteorological instruments.   It is necessary to insist on this, because we often hear reference
made to fishermen, mariners, gardeners, and that class of
persons, as knowing a great deal more about the weather
than other people.   These persons are much abroad in
the open air, and are naturally much interested in
knowing what the weather is likely to be.   But they are
often greatly under the influence of superstitious ideas,
or guided by sayings handed down to them by their

fathers, to which they attach more importance than to anything else. When this is not the case, and their knowledge is really the result of their own observations, they have still nothing but the direction of the wind and the appearances of the sky to guide them in their opinions. It may be sometimes impossible to mistake the indications which these afford; but there are many doubtful states of weather, in which they are wholly insufficient to tell us, without having recourse to the barometer and other instruments, whether it will clear up or rain.

(441.) People, in general, are apt to be especially deceived by bright mornings; whereas the latter are rarely to be trusted, except when the weather is very settled, with the wind in the E. or N.E., and the barometer above 30 inches. It sometimes happens that such a state of things prevails for days, or even weeks together, each morning in succession being bright and cloudless, and the serene sky continuing till night. When this is the case we may safely reckon that "the day will last," as the expression is. But at all other times, and more especially if there has been rain to any amount the day before, however fine it may be in the morning, the sky is almost sure to cloud over as the day advances, with a probability of more rain in the afternoon, if the barometer be low. What tends the more to encourage the fallacious hopes entertained by some persons on these occasions, is the circumstance that the mornings are seldom *so* fine, and the air *so* clear, as during changeable weather when storms are frequent. I well remember, on one occasion, in the old days of travelling by stage-coach, a gentleman, who like myself was a passenger outside, remarking, just as we were setting out under a bright sky, "What a fine day we have for our journey!" On

my replying that it would be necessary to see the day
out before calling it fine, he seemed somewhat surprised;
but he was compelled an hour or two afterwards to own
he had been rather premature in his judgment of the
weather, when clouds began rapidly to form, soon followed by a drenching rain, that speedily brought cloaks
and umbrellas into requisition.

(442.) The case just mentioned is one of those in
which a mere glance at the barometer would reveal what
is likely to happen, notwithstanding the favourable appearances out of doors. Without its help, it is impossible to detect some of the atmospheric conditions, upon
which the weather depends. It is that instrument alone
which tells us what is going on in the upper regions.
If it stand below a certain point, we know that a large
amount of vapour is present in the atmosphere, and that
the weather is not to be trusted, though rain may not
immediately follow. Sometimes, indeed, the barometer
itself may mislead us, if we look only to its absolute
height, and take no account of whether it is rising or
falling. Or, notwithstanding the height of the barometer, there may be a chance of wet, from the great
humidity of the air. Though the actual amount of
vapour in it be inconsiderable, the air may nevertheless
be nearly in a state of saturation, from the lowness of
the temperature. To ascertain what the condition of
the air is, in respect of moisture, it is necessary to call
in the aid of some other instruments besides the barometer. We must have recourse to Daniell's hygrometer,
or to the dry-and-wet bulb thermometer, which will tell
us both the temperature of the air and that of the dew-
point. According to the degree of difference between
these last, we shall be further guided in our judgment

as to the probability of wet; and if an observation of this
kind be combined, as it ought to be, with an observation
of the barometer, we shall then seldom be much at fault;
we shall at least have the best data for determining what
the weather is likely to be, which it is possible to have.

(443.) But the indications afforded by the above instruments will vary with the time of day independently
of the weather. And it will be necessary to bear this
circumstance in mind, when choosing the hours at which
we refer to them. With regard to the barometer, it is
a good practice to consult it in the morning at 9 o'clock,
and, after setting it, to look at it again from time to
time an hour or two afterwards. The natural tendency
of the mercurial column is to fall between 9 A.M. and
3 P.M. If then it be found to rise at that period of the
day, notwithstanding such downward tendency, it is, so
far as it goes, a favourable indication of the weather
being likely to continue fine. Of course the absolute
height of the column must not be left wholly out of the
question; but however low it may have been previously,
the upward movement serves at least to show that the
weather is disposed to mend, rather than otherwise*.

* I had long adopted the above practice before being aware
that Daniell had recommended the same. Speaking of the
hours of 9 A.M., noon, and 3 P.M., as the best hours for making
observations with the barometer, that author says, "Even those
who merely consult the barometer as a weather-glass, would
find it an advantage to attend to these hours; for I have remarked, that much the safest prognostications from this instrument may be derived from observing when the mercury is
inclined to move contrary to its periodical course. If the column
rise between 9 A.M. and 3 P.M., it indicates fine weather; if it
fall from 3 to 9, rain may be expected."—*Meteorological Essays*,
p. 370.

(444.) With regard to the hygrometer and the dry-and-wet bulb thermometer, both the temperature of the air and the temperature of the vapour have a tendency to rise in fair weather, though not in the same ratio, during the first part of the day. We must, therefore, not merely note the difference between these two temperatures, but observe whether this difference increase or diminish, and which it is, whether the dew-point or the temperature of the air, that gains upon the other. If the difference, say at 9 A.M., be already considerable, and this difference is found afterwards to increase, or the temperature of the air to gain upon the dew-point, the barometer at the same time standing high, we have good security for a fine day. If, with a high barometer, the difference be small, and the latter remain nearly stationary, or diminish, not so much owing to the rise of the dew-point, as to the temperature of the air continuing very low, a state of things that often happens in winter, there may be fog or light mizzling rain, but no precipitation to any great amount. If the barometer be low and the dew-point high, and this latter keep gaining upon the temperature as the morning advances, wet may certainly be expected, and perhaps heavy rain of some continuance. If the dew-point and the temperature of the air rise at first in nearly equal proportion, so that the difference between them remains the same, unless this difference be considerable, rain is probable in the afternoon, when " the temperature of the air falls with the declining sun*." And there are other

* Glaisher, whose *Hygrometrical Tables* may be consulted (pp. 14 and 17), for further remarks on the prognostications afforded by the dry-and-wet bulb thermometer. See also Daniell s *Meteorological Essays*, p. 150.

deductions, which a person who is in the habit of observing this instrument along with the barometer will soon learn to make for himself, better than others can make for him.   If at the same time he attend to the direction of the wind and the appearances of the sky, and note the changes which the clouds undergo from time to time, he may not be able to predict the coming weather with absolute certainty, but he will be qualified to give the best opinion about it which it is possible to deliver.

(445.)   For the benefit of those persons who may like to have directions for judging of the weather thrown into the shape of a few compendious rules, I shall conclude this chapter with such, as, on the whole, appear most entitled to confidence.   Some meteorological writers have made large collections of prognostications from different sources, but without sufficient discrimination to be of much use.   People in general would be perplexed by the variety of them, and by finding some more or less contradictory to others.   Thus, Forster, in his "Researches about Atmospherical Phenomena," has devoted a whole chapter to this subject *, which may be consulted by those who wish for more information relating to it than is here given.   To observations of his own, Forster has added a considerable number of proverbs and vulgar sayings about the weather, many of which last, however, are more curious and amusing than useful.

(446.)   We sometimes hear opinions about the season founded on the assumption that, if the weather is of a particular character at any given time, weather of an opposite character may be expected as a compensation

---

* Ch. IV. *Of Indications of Future Changes of Weather.*   See also Lowe's *Prognostications of the Weather, or Signs of Atmospheric Changes*, 1849.

afterwards. And no doubt this is very frequently the case; a period of great wet being often followed by one of drought, or a period in which the temperature is much above the mean, by one in which it falls as much below it. But then the misfortune is, in regard of any expectations of a change at a set time, that we never know how long one of these periods may last. We may fancy that if much rain prevail one month, the next will be fine, or *vice versâ*, and it may be so. But in some instances, the wet, instead of lasting one month, continues two or three, before there is a return to drier weather; or a month's wet may be met in compensation, not by a month's weather of an opposite character, but by weather of an intermediate kind, lasting for a much longer time, and thus restoring the balance more gradually. In fact, when wet or drought is fairly set in, no human sagacity can predict how long it will continue, though we may be often able to read the indications by which the change, when near at hand, is ushered in.

(447.) Little reliance can be placed on any of the prognostics afforded by plants and animals. It has been thought that the latter are affected by approaching changes of weather in different ways; but their movements and behaviour are very uncertain, and often more connected with the season of the year than with any peculiarities of weather. In many cases in which the actions of animals do arise from the influence of the latter, and in which they have been considered as foreboding wet, they are regulated quite as much by temperature and light, as by the approach of rain. Thus, the circumstance of swallows flying low is attendant upon any dull cold weather, when the insects on which they feed, being little on the wing, must be sought for in their retreats in the herbage near the ground. Mr. Lowe has been at the trouble of noting down, in a considerable number of instances, the indications afforded by this class of prognostications, and it will be seen by the results which he has published, that, among those which are usually considered as indications of rain, while some have pointed just as often one way as the other, the greater number have far more often been followed by fine weather*.

(448.) The rules about to be given, and which are the only ones I have ventured to set down in this work, are derived, mainly, from the state of the barometer, in connexion with the wind and the general condition of the atmosphere, or from the appearances of the clouds. They are all either founded upon, or in agreement with, my own observation. Those which have been verbally derived from others, though confirmed by myself, have the authority annexed. What was before stated, however, must be borne in mind, that these rules must be considered, especially when taken singly, as all liable to many exceptions.

Rules for judging of the Weather.

1. The variations of the barometer depend on the variations of the wind. It is highest during frost, with a N.E. wind, and lowest during a thaw, with a S. or S.W. wind.

2. The height of the barometer must be above the mean corresponding to the particular wind blowing at the time, to allow of weather in which any confidence can be placed.

---

* *Prognostications of the Weather*, p. 47.

3. A high and steady barometer is indicative of settled weather.

4. A very low barometer is usually attendant upon stormy weather, with wind and rain at intervals, but the latter not necessarily in any great quantity. If the weather, notwithstanding a very low barometer, is fine and calm, it is not to be depended upon : a change may come on very suddenly.

5. In general the barometer falls before rain; and "all appearances being the same, the higher the barometer, the greater the probability of fair weather."—*Dalton*.

6. If the barometer fall gradually for several days, during the continuance of fine weather, much wet will probably ensue in the end. In like manner, if it keep rising while the wet continues, the weather, after a day or two, is likely to set in fair for some time.

7. Neither a sudden rise nor a sudden fall of the barometer is followed by any lasting change of weather. If the mercury rise and fall by turns, it is indicative of unsettled weather.

8. "If the mercury fall during a high wind from the S.W., S.S.W., or W.S.W., an increasing storm is probable; if the fall be rapid, the wind will be violent, but of short duration; if the fall be slow, the wind will be less violent, but of longer continuance."—*Belville*.

9. "If after a storm of wind and rain, the mercury remain steady at the point to which it had fallen, serene weather may follow without a change of wind; but on the rising of the mercury, rain and a change of wind may be expected."—*Id*.

10. "If the mercury fall with the wind at W., N.W., or N., a great reduction of temperature will follow; in the winter severe frosts, in the summer cold rains."—*Id*.

11. "A steady and considerable fall of the mercury during an E. wind, denotes that the wind will soon go round to the S., unless a heavy fall of snow or rain immediately follow; in this case the *upper* clouds usually come up from the S."—*Belville*.

12. "A fall of the mercury with a S. wind is invariably followed by rain in greater or less quantities."—*Id*.

13. In noticing the wind, regard must be had to whether there are one or more currents in the atmosphere; in the former case the barometer is generally steady and the weather fair, in the latter the mercury fluctuates and the weather is unsettled.

14. A high temperature, with a high dew-point, and the wind S. or S.W., is likely to produce a thunderstorm. If the mercury fall much previous to the storm, the latter is likely to be succeeded by a change of weather. "Sometimes heavy thunder-storms take place overhead without any fall of the mercury; in this case a reduction of temperature does not usually follow."—*Belville*.

15. "A sudden and extreme change of temperature of the atmosphere, either from heat to cold or cold to heat, is generally followed by rain within twenty-four hours."—*Dalton*.

16. "In winter, during a frost, if it begin to snow, the temperature of the air generally rises to 32° (or near it), and continues there whilst the snow falls; after which, if the weather clear up, expect severe cold."—*Id*.

17. A stratus at night, with a generally diffused fog the next morning, is usually followed by a fine day, if the barometer be high and steady. If the barometer keep rising, the fog may last all day; if the barometer be low, the fog will probably turn to rain.

18. Well-defined cumuli, forming a few hours after sunrise, increasing towards the middle of the day, and decreasing towards evening, are indicative of settled weather: if, instead of subsiding in the evening and leaving the sky clear, they keep increasing, they are indicative of wet.

19. A sky dappled with light clouds, of the cirrocumulus form, in the early morning, generally leads to a fine and warm day.

20. A very clear sky, without clouds, is not to be trusted, unless the barometer be high.

21. A sky covered with clouds need not cause apprehension, if the latter are high and of no great density, and the air is still, the barometer at the same time being high. Rain falling under such circumstances is generally light, or of not long continuance.

22. Dark heavy clouds carried rapidly along near the earth are a sign of great disturbance in the atmosphere from conflicting currents: at such times the weather is never settled, and rain extremely probable.

23. After a long run of clear weather, the appearance of light streaks of cirrus cloud at a great elevation is often the first sign of change.

24. Comoid cirri, or cirri " in detached tufts, called ' mares' tails,' may be regarded as a sign of wind, which follows often blowing from the quarter to which the fibrous tails have previously pointed."—*Forster.*

25. Large irregular masses of cloud, " like rocks and towers," are indicative of showery weather: if the barometer be low, rain is all the more probable.

26. The sun setting, after a fine day, behind a heavy bank of clouds, with a falling barometer, is generally indicative of rain or snow according to the season,

either in the night or next morning. In winter, if there has been frost, it is often followed by thaw. Sometimes there will be a rise of temperature only, no rain falling to any amount.

27. A sudden haze coming over the atmosphere is due to the mixing of two currents of unequal temperatures: it may end in rain, or in an increase of temperature; or it may be the precursor of a change, though not immediate.

(449.) The above rules have reference only to the weather a short time in advance of the period of observation. I do not believe that there are any certain rules by which we can predict the character of whole seasons, for reasons already stated (432). Generally speaking, hard winters are preceded by wet and cool summers; and if the wet weather continue through the autumn, there is the more chance of the winter being severe. If, notwithstanding, the weather, during the first two winter months of December and January, be still open, we may expect frost to set in rather severely in February, and to be followed by a cold spring. It has been observed, again, by many, that hot summers are usually preceded by mild winters. And this has certainly been the case in some instances, during the period over which my own observations extend. Still, these rules must be considered as all liable to exceptions. And were it otherwise, in respect of the causes by which the weather is ordinarily determined, it is always possible for we know not what contingencies to arise, to put things out of course, and to give a different turn to the order of phenomena than they would otherwise have taken.

CHAPTER VIII.

OF CLIMATE; MORE PARTICULARLY THAT OF CAMBRIDGESHIRE.

(450.) MANY circumstances combine to give a country that particular climate, by which it is rendered a fit abode for the several plants and animals found in it, as well as adapted to the constitution of man. To judge of its climate, we must take into account both its mean temperature and the relative temperatures of the different seasons,—its humidity, in connexion with the quantity of rain,—its elevation above the sea,—its position in respect of the sea-coast,—the ordinary course of the winds,—as well as the nature of the soil, and the general configuration of the ground. Any alteration in these conditions might materially affect the well-being, and even the existence of many of the creatures inhabiting it. Nor should we overlook the occasional agency of man in bringing about modifications of climate, by such operations as have been alluded to in a former chapter. The changes, indeed, arising from this last source may be confined within narrow limits. Yet they are often sufficient to lead to the utter extirpation of particular species of plants and animals, without any *direct* human interference, while the country in which they take place is improved, and made more healthy, in reference to man himself.

(451.) Of the above points, all needing to be attended to in our inquiries about climate, no doubt the two most important are those which relate to temperature and humidity. There are some mistakes, however, against which we must guard on these heads. Thus, in reference to temperature, it is essential to ascertain its yearly range, and the way in which it varies from one period of the year to another. Two places having the same mean temperature, have not necessarily the same climate as regards heat and cold. The climates of two countries thus circumstanced may be completely opposite, from the different character of the several seasons in the two cases. The summers may be very hot, and the winters severe; or, again, the summers may be cool, and the winters very mild: in the one case, the severity of the winter would tend to lower the mean temperature, as much as the extreme heat of the summer would tend to raise it; in the other, the mildness of the winter would serve to elevate the mean temperature to the same degree that the coolness of the summer would go to depress it. And thus the *mean* temperature of the two countries in question might be identical. Yet how different would be their climates! This difference would at once show itself in the respective produce of the two countries. Many fruits and cereals would ripen in the country enjoying a hot sun and bright skies in summer, notwithstanding the cold of winter, which, in the other, where the skies were constantly clouded, and there was less dissimilarity in the seasons, would be wholly unknown *.

* Humboldt says, that "in no part of the earth, not even in the Canary Islands, in Spain, or in the south of France, has he seen more magnificent fruits, especially grapes, than at Astra-

(452.) The difference just alluded to is that which, generally speaking, exists between an insular and a continental climate, other conditions remaining the same. We see it to a certain extent in the British Islands, which have the heat of summer and the cold of winter both greatly moderated by the waters of the surrounding seas, in comparison of other countries in Europe in the same latitudes. This circumstance tends also to give the weather of our climate that variable character for which it is so remarkably distinguished. On the Continent the seasons are wider apart, and the weather less uncertain. The difference between the mean temperatures of summer and winter in England is several degrees less than on the opposite shores of Holland, and this difference increases the further we advance eastward.

(453.) The immediate cause of this difference is to be sought for in the hygrometric condition of the atmosphere. The country that is surrounded by the sea will have its atmosphere constantly more humid, and the sky more clouded, than another country situate in the same parallel in the interior of a continent. But clouds, it has been shown, in a former part of this work (88), intercept the radiation of heat. The country first mentioned, therefore, will not be so much heated by solar radiation in summer, while it will lose less of its heat by terrestrial radiation in winter; in other words, its

chan, near the shores of the Caspian, in lat. 46° 21′."—Yet the mean annual temperature there is only about 48°, being actually lower than that of Greenwich. But then the mean summer temperature rises to 70°·2, while in winter the thermometer sometimes falls to from —13° to —22°.—*Cosmos* (Sabine's Transl.), vol. i. p. 319.

summers will be cooler, and its winters milder, than those of the other country.

(454.) Again : in reference to the humidity of a climate, we must distinguish between that dampness which is due to an excess of rain, and that which arises from other causes. The amount of vapour contained in the air may, in relation to the temperature, be very considerable, even approaching to saturation, and yet the average quantity of rain be small. It often is so in the neighbourhood of extensive sheets of water, and low marshes that are never dry in consequence of a stiff retentive soil underneath. There is here a constant evaporation going on, supplying moisture to the super-incumbent atmosphere, the effects of which we see in the rich verdure of the pastures adjoining. So too, in all thickly-wooded districts, the humidity of the air is disproportionate to the fall of rain simply considered. Not only is the dampness arising from the latter cause in such places kept up by the shade of the trees cutting off the drying effect of the sun's rays and preventing a circulation of air, but fresh moisture is constantly exhaled by the leaves.

(455.) There are even countries, in which rain is unknown, and which yet are not so dry as to be altogether barren. In the valleys, along the whole coast of Peru, S. of Cape Blanco, it is said, not a drop of rain ever falls ; but for nearly five months in the year, the sky is covered with a kind of fog, which, by its condensation into dew, changes dust into mud, and fertilizes the ground, if not to a sufficient extent for general agricultural purposes, yet at least for the cultivation of particular crops *.

* *Penny Cyclopædia*, Art. " Peru." According to Mr. Darwin, however, the common notion " that rain never falls in the

(456.) Mention has been made above (452.) of the smaller difference between the temperatures of summer and winter in England than on the Continent. The same circumstances that bring this about, give rise to a similar though less considerable difference, in that respect, between some parts of England, which are situate on the south coast, compared with places inland. Proximity to the sea leads here also to cooler summers and milder winters than are experienced in the midland and eastern counties. Hence it is that consumptive patients are sent to winter at Torquay, and other places on the coast of Devonshire, where the comparatively warm and moist air at that season is well suited to cases of delicate lungs.

(457.) Generally speaking, it is a high temperature combined with a humid atmosphere which causes a relaxing climate. But there are mistakes on this subject which are very commonly made, especially in reference to the places just alluded to. Many persons think that because the coast of Devonshire is mild in winter, it must necessarily be very hot in summer. Whereas it is just the contrary, excepting perhaps spots, in which the local configuration of the ground may favour an unusual degree of heat from reflexion, or which are so shut in by hills as to check the circulation of air. That Torquay, in particular, is not the hot and relaxing place in summer that had been supposed, has been clearly shown by Mr. Vivian, in a communication made to the British Association in 1856. From meteorological data

lower part of Peru, can hardly be considered correct," as the "thick drizzling mist," above spoken of, seems, in his estimation, almost to amount to rain, though the quantity that falls, he says, is certainly small.—*Journal*, p. 446.

obtained by that gentleman, it appears not only that the summers of South Devon are rendered cool by the equable temperature of the sea, but that " the humidity of the air in summer is diminished by the same cause, the temperature of the sea being frequently below the dew-point of the air, thus acting as a condenser, and producing results exactly the reverse of the relaxing character assigned to that district by medical and other writers *."

(458.) But, in truth, the terms *bracing* and *relaxing*, as applied to places on the sea-coast, are often very loosely employed. We constantly hear the same place called relaxing by one person, and bracing by another. This shows that neither attaches any very definite meaning to the words. It is one of the many instances in which ideas are caught up from others without inquiry, or adopted by ourselves on insufficient ground. People are much guided in this matter by their own particular feelings, which are so frequently a source of error in meteorology. How common it is to hear persons remark that the day is a little warmer or a little colder than yesterday, simply because it feels so to them, according to their state of health, or according to the temperature of the rooms in which they live, when the thermometer, which is the only true exponent of the real condition of the atmosphere in this respect, indicates just the reverse.

(459.) The public are often similarly deceived in another point. From connecting the idea of a relaxing climate with that of great lassitude and debility, such as is often occasioned by extreme heat, they are apt, forthwith, to pass to the conclusion that a cold bracing

* See *Athenæum* (1856), No. 1504, p. 1057.

air is at all times most conducive to health. It is clear, however, that this must depend greatly upon the period of the year: it may be so in summer, but it is not so in winter, at least in this country. Whether the severe winter cold of such countries as Canada, where the cold is combined with an extremely dry state of the atmosphere, is ordinarily compatible with good health or not, I am not aware. But in Great Britain, surrounded as it is by the sea, the air is probably never sufficiently dry to prevent even the more moderate cold of this climate from having an injurious effect upon the human constitution. Our winters are very variable in character, and it is a vulgar error to suppose that the severe weather we sometimes experience is more healthy than mild *. Such an idea may be prevalent, but it is another instance in which men judge entirely from their own feelings. It is generally from persons in robust health themselves, and with whose particular temperament a cold bracing air best harmonizes, that we hear this remark. They fancy that what suits them will suit all others as well. But how is it in fact? If we turn to the Reports of the Registrar-General, in which we have the best collection of data for determining this question, we find the result going just the opposite way. We there see the mortality in winter bearing a constant ratio to the severity of the weather, and the general

* See, in reference to this subject, a paper by the late Dr. Heberden, entitled "On the Influence of Cold on the Health of the Inhabitants of London," *Phil. Trans.* 1796, p. 279. In this paper the injurious effects of a severe winter on persons generally, but especially on the aged and poor, as well as all whose constitution has been from any cause impaired, is strongly pointed out.

healthiness of the season depending ordinarily upon its mildness. Such at least is the case in the metropolitan district, and there seems no reason why it should be otherwise in the country. If a severe frost set in, the number of deaths in London immediately rises ; when the thaw comes, it immediately falls. And this is not confined, as might be supposed, to cases of death from pulmonary complaints alone, but the same result appears in respect of deaths from numerous other diseases.

(460.) It is foreign, however, to the object of this work to treat generally of climatology, or even of the climate of Great Britain. The remarks that follow refer exclusively to that part of Cambridgeshire in which my own observations have been carried on.

(461.) Cambridgeshire has the character of being damp, and so it undoubtedly is, compared with many other counties in England, notwithstanding the small quantity of rain which falls upon it. But it is in the fens chiefly, or those places, which, like Swaffham Bulbeck, are in the immediate neighbourhood of the fens, that the atmosphere, at certain periods of the year more particularly, is in that humid state which renders it unhealthy for some constitutions. And this circumstance leads to speaking of the soil, which in all countries, as before stated, must necessarily have some influence upon climate. A light sandy soil, through which the rain easily percolates, will, all other circumstances the same, be attended by a drier atmosphere, than a compact clay or peat, in which the wet is retained a long time. A calcareous soil in this respect would hold an intermediate place between the sand and the clay. And in mixed soils, generally, the result will be according to the quality of that material which predominates.

(462.) It is unnecessary here to enter into details connected with the geology of Cambridgeshire, and the origin of the fens, as this subject has been ably treated of by Professor Sedgwick, in a communication to the British Association, to which the reader is referred *. I may simply state that Swaffham Bulbeck is situate on the Lower Chalk, or *Clunch*, as it is provincially called, very near the junction of the lower chalk and galt. The upper greensand, which ordinarily separates these two beds, is reduced in this neighbourhood to a thin stratum, scarcely exceeding a few inches in thick- ness. Very deep graves, occasionally dug in the parish churchyard, show some traces of greensand at bot- tom, by which it is estimated that the latter might be reached at about ten feet. The greensand, however, may be more distinctly traced on the sides of some of the fen-ditches, cropping out from beneath the lower chalk, just at the commencement of the moor, though soon giving way to the galt, upon which the great bed of the fen reposes. The fen itself lies to the N.W., and that portion of it within the parish occupies the entire tract of land between the village and the river Cam. To the E. and N.E. the ground slightly rises, and the lands above the village, in the direction of Newmarket Heath and Burwell, consist wholly of

---

* "On the Geology of the Neighbourhood of Cambridge, including the Formations between the Chalk Escarpment and the Great Bedford Level."—APPENDIX. "Changes in the River Drainage of the Bedford Level, produced by the silting- up of the old water-courses, and the consequent accumulation of Turf-bog and Marsh Lands, &c."—*Report of the 15th Meeting of the British Association, held at Cambridge*, 1845 (Trans. of the Sects.), p. 40.

clunch. These lands are very good for wheat, and that especially grown on the chalk-hills in the vicinity of Burwell is generally more forward than in other places, and the earliest in the market. Much corn is likewise now grown in the reclaimed portions of the fen, but it is liable to be mildewed in wet seasons, and is always comparatively late.

(463.) The time for cutting wheat at Swaffham Bulbeck, on an average of twelve years, I have found to be the 30th of July. But in some hot and dry seasons it has commenced as early as the 16th. The latest period at which I ever knew it begin was the 18th of August. This gives a range of the period for commencing harvest of rather more than a calendar month.

(464.) From the soil in the neighbourhood of Swaffham Bulbeck having in it a large proportion of clay, it is very heavy in winter, and breaks up after a frost, when walked upon, in a very disagreeable manner. As has been observed in the case of similar soils in other places, this last circumstance takes place equally the same, though not to the same degree, whether it has been wet or dry previously to the frost setting in. But familiar as the fact must be to most persons, it has not always been rightly accounted for. Thus White of Selborne has remarked that " if a frost happens, even when the ground is considerably dry, as soon as a thaw takes place, the paths and fields are all in a batter." But I can hardly think that he has hit upon the true cause, when he explains it by saying " that the steam and vapours continually ascending from the earth, are bound in by the frost, and not suffered to escape till released by the thaw*." It is rather the result of moisture

* *Naturalists' Calendar* (by Aikin), p. 141.

absorbed by the ground previous to the frost setting
in ; for it is not observed on all soils. On a sandy soil,
for instance, it is scarcely noticeable ; while on marly
and clayey soils, the quantity of moisture thus absorbed
is very great*. The atmosphere also is in its most
humid state at the time of the minimum night temperature, which, as before shown (213), is mostly regulated by the temperature of the constituent vapour. The
soil, therefore, is enabled to take up the greatest quantity of moisture, just previous to the latter being converted into ice. If, after a frosty night, we break a clod
of earth that has been lying exposed, we find crystals of
ice disseminated through the cracks caused by the ex-
pansion of the freezing moisture. So long as the frost
lasts, the clod still holds together; but when the thaw
comes, the disintegration of the clod causes it to yield
to the slightest pressure, while the moisture it contains
renders it of a pasty consistency. The degree to which
a road or pathway breaks up, when walked upon under
such circumstances, depends upon the looseness of the
surface materials and the severity of the frost. The
first regulates the quantity of moisture imbibed previous

* The late Professor Johnston, in his *Elements of Agricultural
Chemistry and Geology,* p. 104, observes that "during the cooler
season of the night, even when no perceptible dew falls, the
soil has the power of again extracting from the air a portion of
the moisture it had lost during the day. Perfectly pure sand
possesses this power in the least degree ; it absorbs little or no
moisture from the air. A stiff clay, on the other hand, will in
a single night absorb sometimes as much as a thirtieth part of
its own weight, and a dry peat as much as a twelfth of its own
weight; and, generally, the quantity thus drunk in by soils of
various kinds, is dependent upon the proportions of clay and
vegetable matter they severally contain.

to freezing, as the more loosely the materials are laid, the more admission is given to the damp air : the latter regulates the depth to which the expansive power of the frost reaches. A very slight frost is sufficient to make the soil of a ploughed field stick to the shoes in large lumps, while on a hard footpath, and still more on a beaten road, subject to heavier traffic, it requires a frost of some continuance to cause it to break up to the same degree, on the arrival of the thaw.

(465.) There is another circumstance connected with the soil of this part of Cambridgeshire, which I have observed in frosty weather, and which is perhaps worth mentioning. Much of the gravel used in laying down the walks in gardens and shrubberies consists of rounded pebbles of the chalk marl of the neighbourhood, mixed with small pieces of flint. On a frosty morning these chalk pebbles are distinguishable to the eye at a considerable distance by a copious incrustation of ice on their upper surface, often equalling the eighth of an inch in thickness, when there is no such incrustation on the flints similarly exposed. This ice does not appear to be exactly frozen dew, chalk being an indifferent radiator of heat*, and therefore collecting dew but in small quantity. It seems rather to arise from the strong tendency of chalk, like clay, as above mentioned, to absorb moisture from the atmosphere. This moisture, if the temperature of the ground descend to freezing-point, will be congealed on coming into contact with

* Wells found, on exposing at night " clean river-sand, glass, chalk, charcoal, lamp-black, and a brown calx of iron," that chalk produced the least cold of any of those bodies. They were, however, " in the state of a powder, more or less fine," if that makes any difference. See his *Essay on Dew*, p. 171.

the surface of the stone.   As the cooling process gradually extends upwards to the strata of air resting upon the ground, fresh precipitations of moisture take place, and layer is added to layer, until the superficial crust of ice acquires considerable thickness.

(466.) But to return to the subject of climate.   As it is the clayey soil, lying at the bottom of the great level of the fens, that led to the stagnation of the water, by which that portion of the country was originally flooded when the fen was first formed, so the clay may be called the remote cause, or one of the causes, of the dampness of the climate.   Cambridgeshire, however, must have had its climate improved in this respect, from what it formerly was.   The extent to which drainage has been carried on in modern times, especially since the application of steam power to this object, can have hardly failed to render the atmosphere less humid.   Thousands of acres, which were formerly more or less under water, at least during the winter, have been reclaimed in this way, and brought into cultivation, the fertility of the soil being greatly improved by a top-dressing of clay, dug up from underneath the moor.   Crops of oats are usually first sown, and these after a time are succeeded by wheat.

(467.) It would be a singular sight, and to naturalists an interesting one, to see the Cambridgeshire fens as they were in the days of Ray, when presenting only extensive sheets of water, or reedy marshes, inhabited by Cranes and Bitterns, and a vast variety of waterfowl, many of which have now entirely deserted the country, or are only to be met with in unusually wet seasons.   Such a season occurred in the spring of 1824, when a great quantity of rain fell, and vast portions of the fen, which are now ordinarily dry, were completely flooded, and remained throughout the summer in a wetter state than had been remembered for years. Many aquatic birds returned to the fens at that time, and some bred there. Ruffs and reeves, which are now rarely seen, were plentiful, along with godwits, dunlins, ringed plovers, black terns, and other species in less abundance. As a further proof of the altered character of the moor since Ray's time, it may be mentioned that many marsh plants, which would seem to have been not uncommon in Cambridgeshire in his day, are now extremely scarce, if not in some instances extirpated.

(468.) Of course a country that has been thus rendered drier, cannot but have become at the same time more healthy. Diseases of the class of intermittent fevers especially, which are considered as so often taking their origin in marsh miasma, will be greatly lessened when the land is thoroughly drained. One disease in particular I may allude to, viz. ague, which, even within my own time, has almost disappeared in some places, though it is not confidently asserted that the circumstance is due to the above cause. When I commenced residence at Swaffham Bulbeck in 1823, this complaint was so prevalent among the poor of the village, that it was necessary to keep constantly on hand a stock of proper medicine for the relief of those afflicted with it. Of late it has become so infrequent, that hardly more than one or two cases occurred to my knowledge during the last ten years of my living in the neighbourhood.

(469.) It would be a mistake, however, to suppose that the above removal of certain diseases by the drainage of fen-land is owing altogether to the air being thus rendered drier. A damp atmosphere in winter is favour-

able to rheumatic complaints, which are still very prevalent in the low parts of Cambridgeshire at that season.
But to engender fever, so far as it may arise from marsh
miasma, something more than mere humidity of the air,
or even humidity combined with heat, is necessary. It
would seem requisite that there be a certain amount of
decayed vegetable matter upon which these agents can
exert their influence. Local intermittents, therefore,
would be more promoted by the alternate flooding and
drying-up of meadows, according to the season of the
year, than by the land being always under water. In
spots where the water has stagnated in winter, the sun's
influence in summer would thus raise a miasma, the
noxious particles of which might be carried by the winds
to great distances.

(470.) There is still an abundance of decomposing
vegetation in those parts of the Cambridgeshire fens
which have not yet been brought under the plough.
Portions of the fen remain unenclosed, and are pur-
posely kept in a partially undrained state, for the sake
of the peat, or turf, as it is locally called, which is dug
out in large quantities for firing, and to which the poor
in some parishes have a common right. Wherever this
is the case, the moor is broken up into a number of
pits, that speedily become filled with water and aquatic
plants. Among the latter are certain species of *Chara*
and *Potamogeton*, which abound more than any others;
and the first of these especially is well adapted for
causing a rapid accumulation of vegetable remains at
the bottom of the pits, by the constant decaying of the
lower part of its stem, while its upper extremity continues to make fresh shoots. From this circumstance
there is soon formed a decomposing mass, which, as the

heat of summer partially dries up the water, is exposed
to the atmosphere, emitting a most offensive effluvium.
I believe these to be the plants which, during the putrefactive process, mainly contribute to the miasma that
arises from the Cambridgeshire fens. After the accumulation of their remains has proceeded to a certain
extent, the pits are so far lessened in depth that but
little water stands in them in winter, while in summer
they are quite dry. A different kind of vegetation then
takes place. The above plants make way for various
species of *Junci, Carices,* and other grasses, which tend
rapidly to fill the pits up altogether. The decay of these
latter plants in their turn may add to the effect. But
the peculiar strong odour caused by the decomposing
Chara in the deeper pits is very obvious on approaching
one in summer that is half-dried up. If there were a
larger extent of undrained fen than remains at the
present day, and such as existed formerly,—and the
wind during hot weather set in the right direction,—the
noxious exhalation could not fail to affect the atmosphere
of the neighbouring villages, and prejudice the health
of their inhabitants. As it is, the evil is probably yearly
becoming less. All the best turf in Swaffham Fen has
been long since used up. What is now dug there is of
a very inferior quality, and much less profitable for fuel,
and from the extent to which drainage has been carried
on in recent times, the formation of new turf is no
longer possible. While, therefore, the old pits are fast
filling up, comparatively few fresh ones,—which alone
contain a sufficient depth of water to allow of the growth
of the *Chara* and *Potamogeton,*—are excavated. The
time is probably not far distant, when the whole moor
will altogether cease to be worked for turf, and when it

will be found more to the advantage of the owners to have it thoroughly drained and brought into cultivation*.

(471.) Belville, in his *Manual of the Thermometer*†, has given a sketch " of the climate of the eastern parts of England," deduced from observations made at Greenwich, " and extending over a period of thirty-five years," which, in the main, is applicable to Cambridgeshire. So far as there is any peculiarity in the climate of the immediate neighbourhood of Swaffham Bulbeck, it arises from the proximity of the fens, to which attention has been just drawn. This renders the air more damp there than in some places only a short distance off, but which are on slightly higher ground. Thus the air on Newmarket Heath, where there is a much greater depth of chalk above the clay, is perceptibly drier and more invigorating than in the villages between it and the fens, which are lower down, and situate nearer the junction line of those two beds.

(472.) But the greater dampness of the air at Swaffham Bulbeck does not show itself equally all the year round. It is most remarkable in autumn, as might be expected from the increased quantity of rain which falls at that season. And it is, especially during autumn and the early part of winter, when the temperature is on the decline, that I conceive the locality is at all unfavourable to persons in delicate health. The creeping mist, gradually spreading itself over the low meadows, accompanied by a sudden chilliness in the air, as the sun nears the horizon—followed perhaps by a dense

* For further remarks, by the author, " On the Turf of the Cambridgeshire Fens," see *Rep. Brit. Assoc.* 1845 (Trans. of the Sects.), p. 75.   † Page ᴊᴊ.

fog the next day,—is a warning to those who have any tendency to rheumatic or pulmonary diseases, not to expose themselves abroad at such times.

(473.) The summers, however, are quite as hot and dry as elsewhere, and the winters not more severe, than what generally prevail in that part of England. The extreme seasons in one direction are about balanced by those in the opposite. This, however, seems to be the case more with the summers than the winters. Taking the whole period of my residence in the neighbourhood, the hot and cool summers are found to be in equal numbers, but the mild winters to be decidedly in excess of the severe ones.

(474.) From the low situation of Swaffham Bulbeck, frosts are liable to occur rather early in the autumn, showing their destructive effects upon such plants as suffer from a temperature below the freezing-point. During the nineteen years, commencing with 1831, and ending with 1849, the first frost occurred six times in September, eleven times in October, and twice in November. The earliest date of occurrence was September 7th; the latest, November 9th; the mean, October 8th. But I allude here to those frosts only in which the temperature of the air, as indicated by a thermometer at least 5 feet above the ground, has fallen below 32°. Slighter frosts, due simply to terrestrial radiation, the effects of which are confined to a very few inches in height, and which are seldom hurtful to vegetation, are liable to occur, as before stated (93), in almost every month of the year, except July.

(475.) As districts that are low and damp are always subject to frost earlier than others at a greater elevation, so in such places winter often advances more rapidly, though the severity of that season, when fairly set in, may not be greater than elsewhere. Such, at least, seems to be the case at Swaffham Bulbeck, where the temperature, during the last three or four months in the year, declines in a faster ratio than in some parts of England, which are not otherwise different in this respect. It is this circumstance, combined with the damp, which renders the neighbourhood of the fens not a fit place of residence for invalids at that season.

(476.) I had an opportunity of testing the remarkable difference between the temperatures of autumn and winter at Swaffham Bulbeck, and those of the same seasons in one locality on the south coast, which, being much resorted to by patients for the winter half, is worth mentioning. I mean Ventnor in the Isle of Wight. At that place I spent a winter and a spring, and a great part of the previous autumn, the year after I quitted Cambridgeshire. During the period of my stay there, I kept a daily register of the temperature, while there was also one kept at Swaffham Bulbeck by another observer. The difference between the mean temperature of the two places, for the five months of November, December, January, February, and March, I found to be more than $4\frac{1}{2}$ degrees. This, however, was but the result of a single season, which happened to be a very severe one. It will be more instructive, therefore, if we take the mean temperature of each season of the year in succession at the two places, comparing the results obtained at Ventnor by Dr. Martin, from the observations of ten years [*], with those obtained at Swaffham Bulbeck by myself, from the observations

* *The Undercliff,* p. 66.

of nineteen years. These are given in the following Table :—

Mean Temperature.	Ventnor.	Swaffham Bulbeck.	Difference.
Spring	49·82°	47·18°	2·64°
Summer	61·31	60·87	·44
Autumn	53·95	49·86	4·09
Winter	41·80	38·09	3·71

(477.) It is at once seen, on inspecting the above Table, that the difference in the mean temperature of the two places is at a maximum in autumn, after which it keeps diminishing through the winter and spring, until in summer it is reduced to less than half a degree. And this circumstance affords another illustration of what was before observed (457), that places on the sea-coast which are milder in winter than those inland, are not necessarily hotter in summer. Ventnor, sheltered as it is to the North, by the high downs at the back of the town, might be thought by many persons to be very hot in summer, yet we see that, in point of fact, the mean temperature of that season is scarcely higher there than in Cambridgeshire.

(478.) Since the year in which I was in the Isle of Wight, I have resided in the neighbourhood of Bath. And I have thus again had an opportunity of comparing the climates of the West and East of England in one or two respects, which it may not be out of place to allude to in this work. Bath has a higher mean temperature than Swaffham Bulbeck by about a degree and a half, but, as in the case of Ventnor, it is in the winter season that the difference is most striking. I have often noticed that, during severe weather, the ther-

mometer at Swainswick, a village about two miles north
of Bath, has not descended at night within four or five
degrees of the point to which it has fallen in Cambridgeshire, on the same nights, according to the meteorological register kept at the University Observatory,
and published weekly in the *Cambridge Chronicle*. In
summer, on the contrary, during very hot weather, the
temperature has often been higher in Cambridgeshire
than at Swainswick, and never that I am aware lower.

(479). Another circumstance connected with the climate of Bath, which may be worth mentioning, is, that,
so far as my observation has yet gone, the wall-fruit in
my garden at Swainswick does not ripen so early, or
so well, as it did at Swaffham Bulbeck. This may
partly be accounted for by the apparently higher temperature in summer at Swaffham Bulbeck, as just
alluded to. What the real mean summer temperature
at Bath may be, I have not resided long enough in the
neighbourhood to say. Or, perhaps, the above fact may
be explained by the increased quantity of rain which
falls in the West of England compared with what falls
in the East. Wherever there is more rain than at
another place, there will be more cloudiness of the sky,
and less sun; and we know that nothing short of the
direct rays of the sun, shining with little intermission
at the proper season of ripening, will suffice to bring
wall-fruit to perfection. I have no return of the quantity of rain at Bath to offer from my own observations;
but of two parties, from whom I have obtained returns,
one, resident at Radstock Rectory, about eight miles
south of Bath, and 250 feet above the level of the sea,
states 34·64 inches as the average fall in that locality
from ten years' measurements; the other, resident in

Bath itself, gives 32·1 inches as the average there, calculated from the same number of years. It will be seen that the lowest of these returns is nearly half as much again as the average fall of rain at Swaffham Bulbeck, which, as stated in a former part of this work (302), is scarcely 22 inches and a half.

(480.) I will conclude the above slight sketch of the climate of Cambridgeshire, and this whole work, with giving the general character of each successive month in the year at Swaffham Bulbeck, pointing out at the same time any remarkable deviations from that type of weather by which the several months are ordinarily distinguished, that occurred during the period for which my observations were carried on. Likewise, climate being closely connected with vegetation, as shown by the influence which the former exerts over the periodic development of foliage and flowers, in those species of trees, shrubs, and plants, which are either indigenous, or generally cultivated, in any given locality, I will not omit all reference to this part of the subject. It will be sufficient, however, on this occasion, to mention under each month simply those periodic phenomena, which, in respect of the mean date of their occurrence, characterize it the best, and which at the same time, from relating to species for the most part generally diffused in that part of Cambridgeshire, are readily noticed. To the phenomena in the Animal Kingdom I shall make no allusion. The periodic appearances of birds, insects, &c.,—though no doubt dependent to a certain extent on season,—are, nevertheless, neither so closely in connexion with the latter, nor in many cases so regular in the times of their occurrence, as to be of much value in giving to any climate its peculiar stamp ;

add to which, when such phenomena do occur, they may not happen to come under the observer's eye till some time afterwards. Migratory animals are determined in their movements rather by the character of the weather in the country they are leaving, than by that of the country to which they are journeying, or they are under the influence of peculiar instincts, independent of the latter altogether. Hence they often show themselves in certain districts prematurely, when the season, perhaps, is not sufficiently advanced to afford them the requisite nourishment, and either they are obliged to secrete themselves on their arrival till the weather improves, or, if the latter continue severe for any length of time, they die. It is obvious that in such instances the phenomena in question rather mislead, than assist us in our inquiries about climate. Nor is there much less uncertainty in the indication afforded by other species that never quit the neighbourhood in which the observer is situate. A single day of an unusually high temperature in the early part of the year may induce birds to sing, or cause the bat and many insects to come abroad, notwithstanding the season generally may not be particularly forward. Trees and plants, on the contrary, cannot be much forced into leaf and flower, unless weather favourable for a premature development of such phenomena have prevailed for a certain length of time previously. There must be a certain condition of soil and atmospheric influence, in respect of temperature and moisture, necessary for bringing the buds to maturity, which cannot be the effect of two or three warm days only, but must be the result of the whole preceding season. The relative dates, therefore, in different years, at which the buds open, are no bad guides to the character of the season in each particular instance; and the *mean* date will be in accordance with the climate, as judged by its *average* character in the long run. Those who wish to see a full list of periodic phenomena for each month of the year, in both the animal and vegetable kingdoms, may consult the one which I have already given in a former work*.

January.—This month is, on an average, the coldest in the year, as well in respect of the mean temperature, as in respect of the lowest temperatures that occur in this climate. The mean is 36°·88, but it varies in different years from 44°·70 to 29°·30, the variation being greater than that of the mean of any other month. The mean maximum during the day is 40°·8; the mean minimum of the night 32°·7. The mean range of the thermometer is from 51°·04 to 19°·58. The highest temperature ever registered is 57°·5; the lowest 5°. The majority of the nights are more or less frosty.

The month of January was very mild in the following years of the series over which my observations extend; viz. 1834, 1843, 1844, 1845, and 1846, the last four being four years in succession. In 1847, January was very cold, with an unusual prevalence of Easterly winds.

The mean height of the barometer is 29·899 inches. The mean range is 1·484 inches. The full range exceeds that of any other month, the greatest elevations as well as the greatest depressions of the mercurial column both occurring in this. The greatest recorded

---

* "Calendar of Periodic Phenomena in Natural History, as observed in the Neighbourhood of Swaffham Bulbeck," *Observations in Natural History*, pp. 366–412.

height is 30·880 inches; the lowest 28·143 inches; the difference being 2·737 inches.

The prevailing winds for the month are those from the S.-W. and W.-N. quarters, the former being rather in excess of the latter. Sharp weather, with frost of any continuance, is generally the result of the wind passing on to N.-E., and settling itself in that quarter: sometimes, however, severe cold prevails with N.Westerly winds alone, as in January 1841. If the month be an extremely mild one, it is due to the almost exclusive prevalence of Westerly and S.Westerly winds.

The fall of rain and snow is moderate in respect of quantity, though there may be many wet days. The mean fall is 1·452 inches. The quantity never equals that which sometimes occurs in other months; that is to say, the greatest known fall, which is 2·313 inches, is less than that of any other. The least known fall is ·360 inches.

Violent thunder-storms sometimes, though not very often, happen in this, as in the other winter months. Such a storm occurred on the 3rd of January, 1841, in the interval (of a few days only) between two periods of sharp frost, and on the occasion of the wind, which had been previously S.W., shifting rather suddenly to N.W. This January was one of the coldest in the whole series.

The number of periodic phenomena in the vegetable world, speaking of the *Phanerogamia* only, that first show themselves in this month is very limited. The following shrubs and plants, however, if the weather be not very severe, open their flowers in about the following order. Furze (*Ulex europæus*), on a mean of eleven years, on the 13th; Mezereon (*Daphne meze-*

reum), about the 22nd; Hazel (*Corylus avellana*), and Hepatica (*Hepatica triloba*), on a mean of eleven and nine years respectively, on the 25th; Winter Aconite (*Eranthis hiemalis*), on a mean of ten years, on the 26th; Stinking Hellebore (*Helleborus fœtidus*), and Daisy (*Bellis perennis*), on a mean of twelve and eleven years respectively, on the 28th; and the Snowdrop (*Galanthus nivalis*), on a mean of twelve years, on the 30th. These species at least are among those most characteristic of the month. The only shrub that ordinarily comes into leaf so early in the year is the pale perfoliate Honeysuckle (*Lonicera caprifolium*), about the 15th.

February.—A very uncertain month, sometimes mild, with a delicious foretaste of spring, at other times cold, with sharp frosts and cutting winds. The mean temperature is $38°\cdot25$. This slight accession of heat, compared with the mean of January, is almost entirely during the day, the mean maximum rising to $43°\cdot4$, while the minimum of the night, only $33°\cdot1$, is scarcely higher than that of last month. The mean range of the thermometer is from $53°\cdot58$ to $22°\cdot81$. The highest temperature ever registered is $61°\cdot5$; the lowest $13°$. This month was coldest in 1838, when the mean temperature fell to $31°\cdot3$, and warmest in 1846, when it rose to $44°$.

The mean height of the barometer is $29\cdot874$ inches. The mean range is $1\cdot345$ inches. The greatest recorded height is $30\cdot866$ inches; the lowest $28\cdot684$ inches.

Westerly and S.Westerly winds still prevail, as in last month, more than those from any other quarter. But the Northerly, taken as a whole, preponderate over the Southerly, the W.-N. and the N.-E. being in nearly equal proportions. The E.-S. winds which

sometimes occur, would seem to belong to a transitional
state of weather, between the breaking-up of frost and
the return to mild. The S.Westerly winds are attended
by mild and generally wet weather. When N.Easterly
winds prevail, the weather is usually cold and very dry,
with a clear atmosphere, and sharp frosts at night.
If with the same winds there prevail, as sometimes
happens, cold damp weather, with thick mists, and
rain or snow at intervals, it is probably due to an upper
current from the S.W. Such weather occurred in
February 1841.

The quantity of rain that falls this month rather
exceeds, on the average, that of January. The mean
quantity is 1·502 inches. The greatest quantity ever
measured is 3·212 inches; the least is ·270 inches.

The list of plants that ordinarily flower in this
month is somewhat larger than that of the last; still
there are not many, either among the wild or cultivated
species, by which it is particularly characterized, and
which are at the same time of general occurrence. The
following, however, may be mentioned. The Spurge-
Laurel (*Daphne laureola*) flowers, on an average of
twelve years, on the 5th; the Wild Primrose (*Primula
vulgaris*), and the Butcher's Broom (*Ruscus aculeatus*),
during the second week; the Red Dead Nettle (*Lamium
purpureum*), the Dandelion (*Taraxacum officinale*), and
the Spring Crocus (*Crocus vernus*), on an average of
from eight to ten years, during the third; the Lesser
Periwinkle (*Vinca minor*), the Alder and the Yew, each
on an average of twelve years, during the last week.

The leafing season is very little more advanced than
last month. The Elder (*Sambucus nigra*), which is one
of the earliest of shrubs in this respect, comes into leaf,

on an average of twelve years, on the 14th, and the Common Honeysuckle (*Lonicera periclymenum*), on an average of six years, on the 18th. These are the only two species which, in ordinary seasons, leaf in February.

March.—The mean temperature of this month is 41°·7, advancing nearly 3½ degrees upon that of the last. The mean maximum of the day is 48°·2; the mean minimum of the night 35°. The mean range of the thermometer is from 58°·71 to 23°·66. The full range for twenty years, commencing with 1830, and terminating with 1849, is greater than in any other month, the highest temperature recorded being 69°, and the lowest 7°.

The mean height of the barometer is 29·905 inches. The mean range is 1·263 inches. The greatest recorded height is 30·704 inches; the lowest 28·625 inches.

The notoriously cold winds of March are mostly from the N.W. or N.E.; and Northerly winds as a whole prevail over the Southerly in the ratio of six to five. These winds are of great service in drying up the superabundant moisture of the two preceding months, and thereby fitting the ground for the operations of the farmer. The equinoctial gales generally occur some time between the middle and the end of the month, blowing from the S.W. and N.W. by turns, and often breaking up the winter, and introducing milder weather.

The mean rain-fall is 1·633 inches. The greatest quantity ever measured is 2·90 inches; the least ·540 inch.

Judged by its average character, the weather of this month is cold and changeable, with moderate wet, in the form of showers, often of hail or snow, alternating with dry frost. In some years, however, it has very much departed from the type.

Thus in 1835, March was very wet, the fall of rain nearly equalling 3 inches. In 1837 it was very cold, with a mean temperature 5 degrees below the average, and the thermometer in one instance falling to 11°. In 1845 it was still colder than in 1837, with a mean temperature 7 degrees below the average, and only 2°·83 above the freezing-point: on one occasion the thermometer actually fell to 7 degrees. In 1841, on the contrary, March was very fine, with a high mean temperature more than 5 degrees above the average, and the wind generally from the S.W. The last week in March 1830, though previous to the commencement of my series of nineteen years, may be mentioned as unusually hot, the temperature on two days rising to 67° and 69° respectively.

The periodic phenomena of this month, as regards the leafing of trees and shrubs, and the flowering of plants, increase in number considerably. The leafing season, however, can hardly be said to begin, generally, till near the middle of the month,—the first species being the Gooseberry (*Ribes grossularia*), which, taking the mean of twelve years, leafs on the 13th. This is followed by the Syringa (*Philadelphus coronarius*), Lilac (*Syringa vulgaris*), Dog-rose (*Rosa canina*), and Black Currant (*Ribes nigrum*), which come into leaf about the third week; and these by the White-thorn (*Cratægus oxyacantha*), Red Currant (*Ribes rubrum*), Privet (*Ligustrum vulgare*), Bramble (*Rubus fruticosus*), Hazel (*Corylus avellana*), and Mealy-tree (*Viburnum lantana*), which come into leaf during the last week. The Horse-chestnut (*Æsculus hippocastanum*), on an average of twelve years, leafs on the 31st.

The principal plants, whose mean time of flowering is during the present month, are as follows:—Pilewort (*Ranunculus ficaria*), Marsh-marigold (*Caltha palustris*), White Dead-nettle (*Lamium album*), Whitlow-grass (*Draba verna*), and Sweet Violet (*Viola odorata*), flower, on an average of from nine to twelve years, in the course of the first week. The Apricot (*Armeniaca vulgaris*), the Peach (*Persica vulgaris*), Common Coltsfoot (*Tussilago farfara*), and Daffodil (*Narcissus pseudonarcissus*), on a mean of ten or twelve years, during the second week. The Corchorus (*Kerria japonica*), the Aspen (*Populus tremula*), and the Elm (*Ulmus campestris*), the last on a mean of twelve years, during the third week. The White Poplar (*Populus alba*), Dog's Mercury (*Mercurialis perennis*), Gooseberry (*Ribes grossularia*), Cowslip (*Primula veris*), and Hairy Violet (*Viola hirta*), on a mean, varying in the different species from five to eleven years, during the last week.

Oats and barley are generally sown, in the neighbourhood of Swaffham Bulbeck, about the third week of this month,—the sowing of the first preceding by a few days that of the second.

April.—The mean temperature of this month is 45°·88, being about 4 degrees higher than that of March. In both months the accession of heat is still chiefly during the day as in February. The mean maximum of the day is 53°·7; the mean minimum of the night 37°·9. The mean range of the thermometer is from 67°·82 to 27°·86. The highest temperature recorded is 76°·5; the lowest 23°.

The mean height of the barometer is 29·842 inches.

s 2

The mean range is 1·061 inches. The greatest recorded height is 30·526 inches; the lowest 28·965
inches. The same cold winds prevail as in March, the
N.-E. class being at a maximum in this month.

The quantity of rain is rather less than that of either
of the two preceding months, the mean fall being
1·470 inches. The greatest quantity ever measured
was 3·765 inches, in 1846; the least quantity ·210
inch, in 1844.

The weather is generally very changeable this month,
with frequent cold showers from the north-west: there
is often a recurrence of sharp frosts at night, cutting
the early fruits and tender vegetables. Occasionally,
though rarely, very steady, and even hot weather prevails throughout the month. Such was the case in
April 1840, when the mean temperature rose to 48°.
In 1844, April was still more remarkably fine and dry,
with a mean temperature nearly 5 degrees above the
average, the wind being mostly from the S.W. The
country was so parched with drought, that the grass
did not grow, and the farmers in some places were
unable to get their barley into the ground.

The leafing season is now at its height, by far the
greater number of shrubs and trees putting forth their
foliage in this month. Taking the average of a number
of years, varying from five to twelve in different cases,
the following order is observed by the principal species:—
The Peach, Larch, Crab (*Pyrus malus*), Apricot, Cherry,
Barberry (*Berberis vulgaris*), Black-thorn, Common Elm,
and Laburnum (*Cytisus laburnum*), come into leaf during
the first week;—the Pear, Raspberry, Birch, Alder,
Lime, Sycamore, and Wild Guelder-rose (*Viburnum
opulus*), during the second week;—the Maple, Horn-

beam (*Carpinus betulus*), and White Poplar (*Populus alba*), during the third week;—the Beech, Dogwood (*Cornus sanguinea*), Walnut, Wych Elm (*Ulmus montana*), and Ash (*Fraxinus excelsior*), during the last week.

The list of plants that come into flower this month is much extended beyond that of last. Among the species most characteristic of it, and either generally met with, or common in gardens, may be mentioned the Ground-ivy (*Nepeta glechoma*), Box (*Buxus sempervirens*), Crown-Imperial (*Fritillaria imperialis*), Lungwort (*Pulmonaria officinalis*), Plum, and Blackthorn, which generally open during the first week. The Wood Anemone (*Anemone nemorosa*), Laurel (*Prunus laurocerasus*), Pear, Ash, Heart's-ease (*Viola tricolor*), Field Wood-rush (*Luzula campestris*), Red Currant, and Cherry, flower during the second week;—the Strawberry-leaved Cinquefoil (*Potentilla fragariastrum*), Dog-Violet (*Viola canina*), Black Currant, Meadow Lady's-smock (*Cardamine pratensis*), Wild Chervil (*Anthriscus sylvestris*), Fritillary (*Fritillaria meleagris*), Pasque-flower (*Anemone pulsatilla*), and Wood Crowfoot (*Ranunculus auricomus*), during the third week;— the Jack-by-the-hedge (*Alliaria officinalis*), Butter-bur (*Petasites vulgaris*), Buttercup (*Ranunculus bulbosus*), Blue-bell (*Agraphis nutans*), Strawberry (*Fragaria vesca*), Crab, Barberry, Germander Speedwell (*Veronica chamædrys*), Sycamore, Maple, Lilac, and Cuckow-pint (*Arum maculatum*), during the last week.

May.—The weather in this month is usually genial, with the first indications of summer. The mean temperature is 53°·14, being more than 7 degrees above that of April. The mean maximum is 62°·5 ; the mean

minimum 43°·9. The highest temperature recorded
is 83°·5; the lowest 25°. In some seasons May has
been as hot as, if not hotter, than any other month
in the year;—as in 1833, when the mean temperature
rose to 60°·2, being 7 degrees above the average. Also
in the years 1847 and 1848, the mean temperature of
this month was very high. In other seasons, however,
very nipping frosts occur in this month,—sometimes as
late as the last week,—causing great injury to all tender
garden plants and vegetables, as well as to the young
shoots of the forest trees. Among the latter, walnuts,
beeches, and ash, from the circumstance of their leafing
later than other trees, appear to suffer most. Instances
of this unseasonable cold occurred particularly in 1831
and 1838. In the latter year, the thermometer, on the
15th and 16th of this month, fell to 26° and 27°
respectively; and the frost of these two nights, followed by other frosts of less severity, had the effect of
so completely destroying all the young foliage, that the
leaves and shoots, which had been newly put forth,
appeared as if burnt, and, after hanging upon the
branches awhile in a withered state, fell to the ground,
where they were blown about by the wind, and drifted
into heaps, the same as in autumn. The trees, in the
meantime, being thus stripped of their first attire, remained almost naked till midsummer, when new shoots
and foliage were put forth. The woods from this cause
presented the most dreary aspect all the first part of
the summer that I ever witnessed*.

* The frost that occurred in May 1831 was far more severe
in some other parts of England than at Swaffham Bulbeck.
At Edmonton, on the 6th of that month, the temperature during
the night is said to have fallen to 20°, being six degrees lower

In some years, without the occurrence of any frosts of consequence, the mean temperature of May is yet so low as very much to retard vegetation; and when combined with a cold April, the large forest trees have been known to continue bare, as in winter, to quite the end of the month. Such was the case in 1837, in which year the mean temperature of May was only 48°·2.

The mean height of the barometer is 29·958 inches, being higher than the mean of any other month in the year*. The mean range is ·897 inch. The greatest height recorded is 30·572 inches; the lowest is 29·056 inches.

The quantity of rain is 1·768 inches. The greatest quantity ever measured in this month is 4·60 inches; the least quantity ·234 inch. The former occurred in May 1843; the latter in 1848.

The prevailing winds are from the S.-W. and N.-E. quarters, the latter being nearly equal to those of April.

With respect to periodic phenomena, the following species of plants may be selected from among the greatly increased number now in flower, as most regular in their time of opening, and best characterizing this month, when the weather does not run into extremes. The Mealy-tree (*Viburnum lantana*), the Horse-chestnut, Black Medick (*Medicago lupulina*), Bugle (*Ajuga reptans*), Herb-Robert (*Geranium robertianum*), Woodruff (*Asperula odorata*), the White-thorn, Red Clover, and Mountain-ash, on an average of several years, open their flowers during the first week. The Common Fumitory (*Fumaria officinalis*), Bitter Winter-cress

than it had been in the same month, at that place, for forty years.—*Encycl. Metrop.* (Art. Meteorology).

* According to Belville, the highest barometric mean at Greenwich occurs in June.—See *Manual of Barometer*, p. 18.

(*Barbarea vulgaris*), Marsh Valerian (*Valeriana dioica*), Lesser Burnet (*Poterium sanguisorba*), Celandine (*Chelidonium majus*), Milk-wort (*Polygala vulgaris*), Laburnum, Comfrey (*Symphytum officinale*), Lily of the Valley, English Clary (*Salvia verbenacea*), and Columbine (*Aquilegia vulgaris*), flower during the second week. The Star of Bethlehem (*Ornithogalum umbellatum*), Pæony, Solomon's Seal (*Convallaria multiflora*), Upright Crowfoot (*Ranunculus acris*), Twayblade (*Listera ovata*), Monk's-hood (*Aconitum napellus*), Silverweed (*Potentilla anserina*), the Holly, and Ragged Robin (*Lychnis flos-cuculi*), flower during the third week. The Raspberry, White Clover (*Trifolium repens*), White Campion (*Lychnis vespertina*), Herb-Bennet (*Geum urbanum*), Bird's-foot Trefoil (*Lotus corniculatus*), Mouse-ear Hawkweed (*Hieracium pilosella*), Guelderrose, Ox-eye-daisy (*Chrysanthemum leucanthemum*), Hound's-tongue (*Cynoglossum officinale*), Bryony (*Bryonia dioica*), and Syringa (*Philadelphus coronarius*), flower during the last week.

There are a few trees whose mean time of leafing is not till this month. The Fig, the Vine, and the Oak, generally come into leaf just at the beginning of it : the Plane (*Platanus orientalis*) not till the end of the first week ; the Mulberry not till towards the end of the third.

June.—The mean temperature of this month is 59°·25, being an advance upon the last of six degrees. The mean maximum is 68°; the mean minimum 50°·9. The highest temperature recorded is 86°; the lowest 35°. Though the weather in some years is fine and settled throughout, as well as hot, in others it is very changeable, with a great deal of rain in the form of hard showers. It was of the former character in 1845 and 1846, the mean temperature of June 1846 exceeding the average by at least 5 degrees.

Sometimes very cool and cheerless weather occurs, though not particularly wet; as in June 1841, when the mean temperature of eight consecutive days during the first half of the month did not exceed 49°, that of the whole month being 4 degrees below the average. June 1843 was equally cold, with a similarly low mean temperature.

Frosts sometimes occur in this month, as shown by the radiating thermometer placed on or near the ground. Such was the case in 1849, when, on the night of the 13th, the temperature of the grass fell more than two degrees below the freezing-point.

The mean height of the barometer is 29·902 inches. The mean range is ·831 inch. The greatest height recorded is 30·510 inches; the lowest is 29·141 inches.

The mean fall of rain is 2·035 inches. The greatest quantity ever measured is 4·022 inches; the least quantity is 550 inch. The former occurred in June 1841; the latter in 1834.

The prevailing winds are from the S.-W. and W.-N. There is often a constant oscillation of the wind from one to the other of these two quarters, especially during changeable weather, as described in a former part of this work (396).

A violent storm of hail and wind occurred on the 5th of June, 1831; and an extremely violent wind on the 11th of June, 1833.

The periodic phenomena of this month in the vegetable kingdom are extremely numerous, it being the chief flowering month for plants. The following species, most of them very common, best mark the progress of the

month, by the order in which they successively open
their flowers :—The Elder (*Sambucus nigra*), on an
average of twelve years, flowers on the 1st day of the
month. The Red Poppy (*Papaver rhœas*), Cow-parsnep
(*Heracleum sphondylium*), Rye, Deadly Nightshade
(*Atropa belladonna*), Spotted Orchis (*Orchis maculata*),
Bladder Campion (*Silene inflata*), and Saintfoin (*Ono-
brychis sativa*), flower in the course of the first week.
The Common Honeysuckle (*Lonicera periclymenum*),
Common Vetch (*Vicia sativa*), Dogwood (*Cornus san-
guinea*), Small Bindweed (*Convolvulus arvensis*), Dog-
Rose (*Rosa canina*), Yellow Flag (*Iris pseudacorus*),
Mallow (*Malva sylvestris*), Wild Thyme (*Thymus ser-
pyllum*), Great Snapdragon (*Antirrhinum majus*), Hedge
Woundwort (*Stachys sylvatica*), Great Hedge Bedstraw
(*Galium mollugo*), flower in the course of the second
week.   Black Knapweed (*Centaurea nigra*), Meadow
Crane's-bill (*Geranium pratense*), Corn-cockle (*Lychnis
githago*), Water Betony (*Scrophularia aquatica*), Bee-
Orchis (*Ophrys apifera*), Foxglove (*Digitalis purpurea*),
and Stone-crop (*Sedum acre*), flower during the third
week.   Wheat, Stinking Horehound (*Ballota nigra*),
French Willow-herb (*Epilobium angustifolium*), Privet,
Field Scabious (*Knautia arvensis*), Round-leaved Bell-
flower (*Campanula rotundifolia*), Millefoil (*Achillea mil-
lefolium*), Meadow-sweet (*Spiræa ulmaria*), and Agri-
mony (*Agrimonia eupatoria*), flower during the last week.
    Meadow hay is cut at Swaffham Bulbeck, on an
average of twelve years, on the 11th of June.   Strawberries ripen, on an average, on the 22nd, and Cherries
a few days later.
    July.—This is on the whole the hottest month in the
year, though the difference between it and August is but

trifling. The mean temperature is 62°·2, being about 3 degrees higher than that of June. The mean daily maximum is 70°·8; the mean minimum of the night 53°·2. This latter is higher than that of any other month, excepting August, the warmest nights occurring in these two months: consequently, at this season the increase of the mean temperature is derived mainly from that of the night. The mean range of the thermometer is from 80°·61 to 42°·79. The highest temperature recorded is 93°; the lowest 36°. July 1846 and July 1847 were both very hot, the mean temperature of the former exceeding the average by 4 degrees. July 1840 and July 1841 were both very cold, the mean temperature in each instance being 3 degrees or more below the average. The weather in this month is more often wet or changeable, than fine or settled: when changeable, with a high temperature, thunder-storms often ensue.

The mean height of the barometer is 29·934 inches. The mean range is ·752 inch, being less than in any other month of the year. The greatest height recorded is 30·400 inches; the lowest is 29·205 inches.

The mean fall of rain is 2·008 inches. The greatest quantity ever measured is 3·720 inches; the least quantity is ·680 inch.

The prevailing winds are from the N.W., but the S.Westerly are likewise considerable, especially in rainy seasons. The wind is much disposed at such times to oscillate between the S.-W. and W.-N. quarters, as in last month; sometimes running as far each way as S. and N. respectively. When rain occurs from the N.W. or N., it is principally in the form of hard showers or storms: continued wet from the same points is attended either by a low barometer or a low temperature.

The periodic phenomena of this month are as numerous as in the last. The flowering of plants, which keeps continually increasing from the commencement of the year, now attains its maximum. I confine myself to mentioning the following species, as some of those which are most characteristic of the month, and which ordinarily flower in the order in which they are here given. Blackberry (*Rubus fruticosus*), Lime (*Tilia europæa*), Ragwort (*Senecio jacobæa*), St. John's-wort (*Hypericum perforatum*), White Lily (*Lilium candidum*), and White Jasmine (*Jasminum officinale*), flower during the first week. Everlasting Pea (*Lathyrus latifolius*), Great Bindweed (*Convolvulus sepium*), Succory (*Cichorium intybus*), Cat-mint (*Nepeta cataria*), Purple Loosestrife (*Lythrum salicaria*), Traveller's Joy (*Clematis vitalba*), Vervain (*Verbena officinalis*), and Large-flowered St. John's-wort (*Hypericum calycinum*), during the second week. Great Willow-herb (*Epilobium hirsutum*), White Poppy (*Papaver somniferum*), Eye-bright (*Euphrasia officinalis*), Tutsan (*Hypericum androsæmum*), Cotton-thistle (*Onopordum acanthium*), Square-stalked St. John's-wort (*Hypericum quadrangulum*), Spear-thistle (*Carduus lanceolatus*), and Small Teasel (*Dipsacus pilosus*), during the third week. Hemp-agrimony (*Eupatorium cannabinum*), Flea-bane (*Pulicaria dysenterica*), Burdock (*Arctium lappa*), Wild Teasel (*Dipsacus sylvestris*), Hairy Mint (*Mentha sativa*), Calamint (*Calamintha officinalis*), and Gipsy-wort (*Lycopus europæus*), during the last week.

Currants, Raspberries, and Gooseberries, ripen at Swaffham Bulbeck on an average of years, during the first week of this month; Apricots in the last week.

August.—This month in the long run partakes very much of the same character as the last. The mean temperature is 61°·74; the mean maximum of the day 70°; and the minimum of the night 53°·2, this last being the same as that of July. The mean range of the thermometer is from 79°·47 to 43°·13. The highest temperature recorded is 88°; the lowest 38°.

In 1842 and 1846, this month was remarkable for its high temperature, exceeding the average, in the former of these two years, by 5 degrees. In 1844 and 1845, August was cold and changeable, with much wet; in the latter year, the temperature was 4 degrees below the average, and the harvest very backward, not commencing till the 18th; this was also the season of the first appearance of the potatoe-disease. Frosts very rarely, though they do sometimes, occur in this month (93).

The mean height of the barometer is 29·872 inches. The mean range is 863 inch. The greatest height recorded is 30·371 inches, being exceeded by that of every other month in the year. The lowest recorded is 29·106 inches. The winds are much the same as in the last month, or chiefly from West to North.

The quantity of rain that falls in this month is very variable in different years, more so, according to my own observations of seventeen years, than in any other. Exceedingly heavy storms of rain and hail sometimes occur, as during the awful tempest that did so much damage in the neighbourhood of Cambridge in August 1843, before spoken of (377). August 1833 was likewise remarkable for a great fall of rain, amounting to more than 2 inches, that took place on the last two days of the month, accompanied by a hurricane of wind (319). In other seasons, however, the weather is steady throughout, and very little rain falls at all. The mean quantity is 2·475 inches. The greatest quantity ever measured is 5·765 inches; the least quantity ·210 inch.

The periodic phenomena of this month, connected with the vegetable world, are few compared with those of June and July. The flowering season of plants is on the decline, and there is but a limited number of indigenous species, which now open for the first time. Among these the following may be enumerated :—Wild Angelica (*Angelica sylvestris*), Wormwood (*Artemisia absinthium*), Large-flowered Hemp-nettle (*Galeopsis versicolor*), Mugwort (*Artemisia vulgaris*), Carline-thistle (*Carlina vulgaris*), Purple Melic-grass (*Molinia cærulea*), Red Goosefoot (*Chenopodium rubrum*), Soapwort (*Saponaria officinalis*), Devil's-bit Scabious (*Scabiosa succisa*), Tansy (*Tanacetum vulgare*), Woolly-headed Thistle (*Carduus eriophorus*), and Autumnal Gentian (*Gentiana amarella*). These furnish a succession of species, the first three or four flowering, on an average of years, quite at the commencement of the month, the last two seldom much before the end of it. Peaches ripen, on an average, at Swaffham Bulbeck on the 31st.

September.—Belville remarks that this month "wavers between summer and autumn." Occasionally hot summer weather is protracted quite to the end of it, as in 1832, when the thermometer rose above 70° for six successive days during the last week. In 1842, also, the temperature was 4 degrees above the mean, this latter being 56°·94. In 1845 and 1847, on the other hand, September was very cold, with an average temperature more than 4° below the mean.

Frosts—speaking now of cases in which the temperature of the air, to the height of at least 10 feet above the ground, falls below the freezing-point—are not unfrequent at Swaffham Bulbeck in this month, though more often they do not occur till the next. The mean maximum temperature of the day is 65°·2 ; the mean minimum 49°·5. The mean range of the thermometer is from 74°·53 to 36°·35. The highest temperature ever registered is 80°·5 ; the lowest 28°.

The mean height of the barometer is 29·899 inches. The mean range is ·892 inch. The greatest height recorded is 30·541 inches ; the lowest 28·990 inches. The winds are much as in August and July, except that generally towards the end of the month, the equinoctial gales occur, blowing mostly from the W. or S.W.

The average quantity of rain is 2·164 inches. The greatest quantity ever measured in this month is 3·926 inches ; the least quantity ·271 inch.

The number of plants that first flower in this month is greatly reduced. The Meadow-saffron (*Colchicum autumnale*) flowers in gardens early in the first week : the Ivy, which is most characteristic of the month, flowers at Swaffham Bulbeck, on an average of eleven years, on the 26th. The Laurestine (*Viburnum tinus*) also generally during the last week. The approach of autumn is indicated by the commencement of the fall of the leaf, in the Lime, Wych Elm and Sycamore, about the third week.

October.—The mean temperature of this month, which falls nearly 7° below that of September, is 50°·17, approaching more nearly to the mean of the whole year than that of any other month. The mean maximum of the day is 56°·2 ; the mean minimum of the night 43°·9. The mean range of the thermometer is from 66°·08 to 31°·42. The highest temperature ever registered is 74° ; the lowest 26°. October 1831 was very warm with a mean temperature of 56°·2, nearly equalling that of September, and with only one night of very slight frost. In each of the two years 1835 and 1842, on the contrary, October was very cold, with a mean temperature of only 46°, more than 10° less than in 1831, and attended in 1842 by a great prevalence of northerly winds.

The mean height of the barometer for October is 29·847 inches. The mean range is 1·335 inches. The greatest height recorded is 30·718 inches; the lowest 28·770 inches. The winds are variable, but those from the S.–W. quarter rather preponderate.

October is in general the wettest month in the year, the mean fall of rain, which equals 2·575 inches, exceeding that of any other. The wet is often excessive, amounting to more than 4 inches, and much interfering with the sowing of wheat, which, though usually done in Cambridgeshire in this month, in some seasons it has been found necessary to defer, in consequence of the continued rains, to the beginning of November.

The greatest quantity of rain ever measured was in October 1841, when it amounted to 4·383 inches, coming almost entirely from the S.W. The quantity was nearly as great in 1843 and 1844, both years more than 4 inches; but there was this difference between them, that, while in the latter year the rain was principally from the S.W., with a mean temperature of nearly 50°, in 1843, on the contrary, there was a prevalence of northerly winds, and the mean temperature only 47°·5.

The driest October in my series of years was that of 1834, when the rain amounted to scarcely more than an inch. October 1845 was also very dry, the quantity being only 1·101 inches, and hardly any falling after the first ten days: the ground became hard, and the roads dusty, as in the middle of summer, and farmers could not, in this exceptional instance, prepare their land for wheat-sowing on account of the drought.

The last week in October 1836 was remarkable for its severity, there being sharp frosts at night, and a heavy drifting fall of snow, which lay on the ground to the depth of several inches for days afterwards. I never remember any such snow as this so early in the season before. Newmarket Heath was so covered with snow (it being the time of the races) that it was actually found necessary to sweep the course before the horses could run. A great deal of fruit was still ungathered, and the apples in many places were hanging upon the trees, coated with snow.

The flowering season—as regards those species of indigenous plants that naturally first open in October— is perhaps at its minimum in this month. Among the phanerogamous plants, which alone I here speak of, there is scarce any to be noticed, except the Strawberry-tree (*Arbutus unedo*), (and that of course in Cambridge-shire only in shrubberies and gardens) flowering, on an average, about the second week. The chief periodic phenomena characteristic of the month, in the vegetable world, are the altered tints of the foliage of forest trees, and the fall of the leaf, the latter becoming more and more general as the autumn advances. The Horse-chestnut, Walnut, Birch and Maple, begin generally to shed their leaves in the first week; the Beech, White Poplar, Cherry, Ash, Aspen, and Lombardy Poplar, during the second; the Hazel, Virginian Creeper, and White-thorn during the third; the Elm (old trees at least), on an average of years, not till quite the end of the month. The Lime, which is earlier in losing its leaves than most other trees, is stripped in the avenue at Bottisham Hall, taking the mean of seven years, on the 18th. Wheat-sowing begins, on the same average, on the 7th.

November.—The mean temperature of this month, which has at least an equal fall to that of September, is 42°·67. It varies less from year to year, or has a *less range* than that of any other month. The mean maximum of the day is 48°; the mean minimum of the night 37°·4. The mean range of the thermometer is from 57°·92 to 26°·79. The highest temperature ever registered is 62°; the lowest 24°. Frost is seldom severe or continues for more than a few nights together.

The mean height of the barometer is 29·803 inches, and descends to a minimum in this month. The mean range is 1·349 inches. The greatest elevation recorded is 30·587 inches; the lowest 28·510 inches. S.Westerly winds prevail, and often blow with considerable violence; increasing, in respect of frequency, upon those of October.

The weather is usually changeable and unsettled, with much rain, the atmosphere very damp and often nearly saturated with moisture. From this latter circumstance thick fogs are frequent, upon any sudden shift of the wind to a colder quarter. The mean rain-fall is 1·903 inches. The greatest quantity ever registered is 3·645 inches; the least quantity ·921 inch.

The instances most exceptional to the ordinary character of this month, were those of November 1845, 1846 and 1847, three Novembers in succession, which were all remarkable for their high temperature and fine steady weather prevailing for the greater part of the month. The first two of them were both very dry, that of 1846 especially so, when S.Easterly winds predominated, and the roads were dusty as in March to a late period. Also from the great drought, the leaves continued upon the trees for a considerable time after they had withered,—not falling to the ground immediately, as ordinarily takes place, the moisture being insufficient to induce decay and separation of the leaf-stalks from the branches. This gave the trees a very unusual appearance. The mean temperature of this month (Nov. 1846) was 45°·5, being about 3° above the average.

The flowering season is quite at an end in this month, and the successive stripping of the trees and shrubs is the only periodic phenomenon in the vegetable kingdom. The Elm and Oak are the last to become bare, and these having lost all their leaves, the trees are everywhere stripped, on an average of years, about the last day of the month.

December —In the long run, the weather of this month is much oftener mild than severe in Cambridgeshire. The mean temperature is 39°·19, its fall being scarcely half that of either of the two preceding months. The mean maximum of the day is 43°·5 ; the mean minimum of the night 35°·1. The mean range of the thermometer is from 54°·4 to 24°·23. The highest temperature ever registered is 58°* ; the lowest 15°. Very sharp frosts, or lasting for more than two or three nights together, are the exception—not the rule. On the average, half, or rather more than half, the nights in the month are frosty.

* On December 13, 1842, and also on December 23, 1843, the thermometer rose to 60° at the Cambridge Observatory. (*Camb. Chron.* Dec. 13, 1856.)

Out of nineteen Decembers, eleven have been mild, with a mean temperature varying from one to six degrees above the average; eight cold, or with a mean temperature below the average, but only three very severe. The mildest were those of 1833, 1842, and 1843, in each of which the mean temperature was nearly 45°*, attended by Westerly and S.Westerly winds, with scarce any frost, and not much wet. In the latter two instances the weather was remarkably fine as well as mild, and on Dec. 24th, 1843, butterflies were on wing, and birds singing as in spring.

The severest Decembers were those of 1840, 1844, and 1846. In the former two the mean temperature of the month was hardly above freezing-point, and in 1846 it fell as low as 31°. In all these Decembers Easterly and N.Easterly winds prevailed, and sharp frost was the result.

The mean height of the barometer for December is 29·924 inches. The mean range is 1·345 inches. The greatest elevation recorded is 30·722 inches; the lowest 28·388 inches.

Taking one year with another, Westerly and S.Westerly winds predominate. But notwithstanding the decided prevalence of these winds in ordinary seasons, the mean fall of rain at Swaffham Bulbeck would seem by my measurements to be less in December than that of any other month, being only 1·364 inches. This may per-

* The mean temperature of December last (1857) at Upper Swainswick, near Bath, was 46°·2, and the mildest within my experience. A thermometer 4 feet from the ground did not fall below 40° more than six times in the whole month, and not until the night of the 30th below freezing-point, the minimum then being only 31°.

haps arise from the circumstance of these winds having already parted with their great excess of moisture during the rains of the two months preceding. The greatest quantity of rain ever measured is 2·903 inches in 1845; the least quantity ·183 inch in 1843. This latter is the smallest quantity of rain ever registered by me in any month during the seventeen years for which I measured it.

Heavy falls of snow are not usual in Cambridgeshire, and when they do occur, seldom happen in this month; the only remarkable instance to the contrary was the drifting snow on Christmas-day and the day following in December 1836, before alluded to (344).

Occasionally in this month, as well as in January, dull monotonous weather prevails for days together, with a N.Easterly wind, and a sky thickly clouded, the sun never showing itself; the atmosphere is damp and *raw*, though there is little or no actual rain ;—the temperature very stationary, low, but not falling, or only just falling, to the freezing-point. It is during such unhealthy weather that influenza, rheumatism and lumbago frequently occur; as in December 1847, when this kind of weather continued for ten days without variation,—and previously, in January 1837, when influenza was general throughout the country to an almost unprecedented degree.

There are no periodic phenomena, particularly connected with this month, in that part of England to which most of the observations contained in this work relate. In very mild winters some few plants, which do not ordinarily flower before January, open in December; but, generally speaking, the vegetable world, exclusive of Cryptogams, lies now dormant, and to all outward appearance dead, though yet in secret collecting its vital energies anew, preparatory to again displaying in succession, on the return of the seasons, those numerous species, which characterize the successive months of each year, and which, by the order and regularity they observe in the times of their appearance, no less than by the variety of their forms and colours, excite the admiration of every reflecting mind, leading it up to Him, who is the first source of Nature's choicest gifts, and whose watchful Providence "upholds all things" in their courses "by the Word of his Power."

"WHILE THE EARTH REMAINETH, SEED-TIME AND HARVEST, AND COLD AND HEAT, AND SUMMER AND WINTER, AND DAY AND NIGHT SHALL NOT CEASE*."

* *Gen.* viii. v. 22.

GENERAL INDEX.

with fine and wet weather
explained, 134, 135.

Barometer, rise of, sometimes
very rapid, 148.

——, sometimes stationary for a
long time together, 148.

——, state of, during rain, 156.

——, table showing mean height
of, at Swaffham Bulbeck
during each month of the
year, 149.

——, table showing mean yearly
height and yearly ex-
tremes of, at Swaffham
Bulbeck from 1831 to
1849, 137.

——, the words *Fair*, *Change*,
and *Rain*, not to be relied
on, 130, 159.

——, use and importance of, as
a weather-glass, 129.

Blue mist, 222.

C.

Chalk - pebbles in gravel - walks
become incrusted with ice on
frosty mornings, 369.

Chara, offensive effluvium of, a
probable cause of miasma, 373.

Clearing of forests and drainage
of marshes, their effect on climate, 343.

Climate, circumstances which influence, 358, 361.

——, humidity of, not necessarily
connected with the quan-
tity of rain, 162.

——, mistakes respecting temperature and humidity to
be guarded against in our
inquiries respecting, 359,
361.

Climate, much determined by
the quantity of vapour
in the air, 165.

—— of Cambridgeshire, probably
improved from what it
formerly was, 370.

—— of England, probably not
changed, 5.

——, the terms "bracing" and
"relaxing" as applied to,
loosely employed, 363.

——, what causes a relaxing, 362.

Climates, difference between insular and continental, 360.

——, difference between maritime
and inland, 362.

—— may be very different,though
having the same mean
temperature, 359.

Cloud, definition of a, 189.

Clouds, different, have different
elevations, 190.

——, effect of, in checking solar
and terrestrial radiation,
63.

—— on the summits of mountains apparently station-
ary, 196.

——, principal forms of, 191.

——, the so-called aqueous vesicles of which they consist,
189.

——, two strata of, one above the
other, during rain, 195.

Cold, degree of, much affected by
local circumstances, 10.

——, extremes of, in what months
they occur, 14.

——, great differences sometimes
in the degree at two
contiguous places, 11.

——, ——, how explained, 12.

THE END.

PRINTED BY TAYLOR AND FRANCIS,
RED LION COURT, FLEET STREET.

Printed in the United States
By Bookmasters